ESSENTIALS OF BIOPHYSICS

Edited by
Deah E. Elmundo

Delve Publishing

Essentials of Biophysics

Edited by Deah E. Elmundo

Edition 2018

Delve Publishing

Published by
Delve Publishing LLC
708 3rd Avenue, 6th Floor
New York
NY 10801
United States of America
www.delvepublishing.com

ISBN: 978-1-68095-839-3 (Hardback)

Printed in United States of America

Library of Congress Control Number: 2016920685

About the Editor

Deah E. Elmundo

Deah obtained her BS in Physics for Teachers from the Philippine Normal University where she graduated Magna cum laude and completed her academic requirements leading to MS in Teaching Physics from De La Salle University. She loves teaching Classical and Modern Physics and is presently a full-time science faculty of a junior high school in her country. She is also a maths and physics content specialist to a few tuition companies in Singapore and Australia

Table of Contents

3 CARBOHYDRATES BIOMOLECULES

4 MEMBRANE LIPIDS

5 ENZYME KINETICS

6 STUCTURAL DETERMINATION TECHNIQUES

7 APPLIED PHYSICS IN MEDICINE

8 BIOENERGETICS

9 BIOLOGICAL SYSTEM

List of Abbreviations

LT	Leukotrienes
RHEED	Reflection High Energy Electron Diffraction
SEM	Scanning Electron Microscope
SAXS	Small-Angle X-Ray Scattering
SPLEEM	Spin-Polarized Low-Energy Electron Microscopy
STED	Stimulated Emission Depletion
MGDG	Monogalactosyl Diglyceride
MPIO	Micron-Sized Particles Of Iron Oxide
NIR	Near Infrared
TNA	Threose Nucleic Acid
TIRF	Total Internal Reflection Fluorescence
TRNA	Transfer Rna
TEM	Transmission
UBP	Unnatural Base Pair
UBP	Unnatural Base Pair
PAL	Physical Activity Level
PUFAS	Polyunsaturated Fatty Acids
PDB	Protein Data Bank
PBGS	Porphobilinogen Synthase

Preface

Biophysics is that branch of knowledge that applies the principles of physics and chemistry and the methods of mathematical analysis and computer modeling to understand how biological systems work. Biophysics is a bridge between biology and physics. Biology studies life in its variety and complexity. It describes how organisms go about getting food, communicating, sensing the environment, and reproducing. On the other hand, physics looks for mathematical laws of nature and makes detailed predictions about the forces that drive idealized systems. Spanning the distance between the complexity of life and the simplicity of physical laws is the challenge of biophysics. Looking for the patterns in life and analyzing them with math and physics is a powerful way to gain insights. Biophysics can trace its roots back to the ancient Greeks and Romans, who first developed hypotheses about the physical basis for consciousness and perception. It underwent rapid development after World War II, stimulated in part by the application of nuclear physics to biological systems, including the investigation of radiation effects on living matter. In the course of these studies, physicists were introduced to biologists and biological problems, and biophysics evolved as a new scientific field. An important area of biophysical study is the detailed analysis of the structure of molecules in living systems. The best-known achievement related to this area is the model of deoxyribonucleic acid (DNA), the hereditary material of life. Biophysics is a wellspring of innovation for our high-tech economy. The applications of biophysics depend on society's needs. In the 20th century, great progress was made in treating disease. Biophysics helped create powerful vaccines against infectious diseases. It described and controlled diseases of metabolism, such as diabetes. And biophysics provided both the tools and the understanding for treating the diseases of growth known as cancers. Today we are learning more about the biology of health and society is deeply concerned about the health of our planet. Biophysical methods are increasingly used to serve everyday needs, from forensic science to bioremediation. Biophysics gives us medical imaging technologies including MRI, CAT scans, PET scans, and sonograms for diagnosing diseases. It provides the life-saving treatment methods of kidney dialysis, radiation therapy, cardiac defibrillators, and pacemakers. Biophysicists invented instruments for detecting, purifying, imaging, and manipulating chemicals and materials. Biophysics discovers how to modify microorganisms for biofuel (replacing gasoline and diesel fuel) and bioelectricity (replacing petroleum products and coal for producing electricity). Biophysics discovers the biological cycles of heat, light, water, carbon, nitrogen, oxygen, heat, and organisms throughout our planet. Biophysics harnesses microorganisms to clean our water and to produce lifesaving drugs.

Content Coverage

Chapters One and *Two* focus on protein biomolecules and news journalism, respectively. Proteins perform a vast array of functions within organisms, including catalysing metabolic reactions, DNA replication, responding to stimuli, and transporting molecules from one location to another, whereas nucleic acids are biopolymers, or large biomolecules, essential for all known forms of life.

Chapters Three and *Four* present coverage about carbohydrates and membrane lipids, separately. Carbohydrates perform numerous roles in living organisms, which provide energy for humans and animals, while Membrane lipids are a group of compounds which form the double-layered surface of all cells.

Chapters Five and *Six* focus on the enzyme kinetics and stuctural determination techniques. Enzyme kinetics is the study of the chemical reactions that are catalyzed by enzymes, whereas most commonly structure determination techniques are X-ray crystallography, NMR spectroscopy, electron microscopy and molecular modelling.

Chapter Seven and *Eight* presents about biomedical physics and bioenergetics, distinctly. Biomedical physics involves diagnosing and treating diseases, whereas bioenergetics includes the study of the transformation of energy in living organisms and the study of thousands of different cellular processes such as cellular respiration and the many other metabolic processes that can lead to production and utilization of energy in forms such as ATP molecules.

Chapter Nine covers about biological system; a complex network of biologically relevant entities As biological organization spans several scales, examples of biological systems are populations of organisms, or on the organ- and tissue scale in mammals and other animals, the circulatory system, the respiratory system, the nervous system, etc.

Protein Biomolecules

Chapter Contents

Protein is a macronutrient that is essential to building muscle mass. It is commonly found in animal products, though is also present in other sources, such as nuts and legumes.

There are three macronutrients: protein, fats and carbohydrates. Macronutrients provide calories, or energy. The body requires large amounts of macronutrients to sustain life, hence the term "macro," according to the University of Illinois McKinley Health Center. Each gram of protein contains 4 calories. Protein makes up about 15 percent of a person's body weight.

INTRODUCTION

Proteins are large biomolecules, or macromolecules, consisting of one or more long chains of amino acid residues. Proteins perform a vast array of functions within organisms, including catalysing metabolic reactions, DNA replication, responding to stimuli, and transporting molecules from one location to another. Proteins differ from one another primarily in their sequence of amino acids, which is dictated by the nucleotide sequence of their genes, and which usually results in protein folding into a specific three-dimensional structure that determines its activity. A linear chain of amino acid residues is called a polypeptide. A protein contains at least one long polypeptide. Short polypeptides, containing less than

20–30 residues, are rarely considered to be proteins and are commonly called peptides, or sometimes oligopeptides. The individual amino acid residues are bonded together by peptide bonds and adjacent amino acid residues. The sequence of amino acid residues in a protein is defined by the sequence of a gene, which is encoded in the genetic code. In general, the genetic code specifies 20 standard amino acids; however, in certain organisms the genetic code can include selenocysteine and—in certain archaea—pyrrolysine. Shortly after or even during synthesis, the residues in a protein are often chemically modified by post-translational modification, which alters the physical and chemical properties, folding, stability, activity, and ultimately, the function of the proteins. Sometimes proteins have non-peptide groups attached, which can be called prosthetic groups or cofactors. Proteins can also work together to achieve a particular function, and they often associate to form stable protein complexes.

Once formed, proteins only exist for a certain period of time and are then degraded and recycled by the cell's machinery through the process of protein turnover. A protein's lifespan is measured in terms of its half-life and covers a wide range. They can exist for minutes or years with an average lifespan of 1–2 days in mammalian cells. Abnormal and or misfolded proteins are degraded more rapidly either due to being targeted for destruction or due to being unstable.

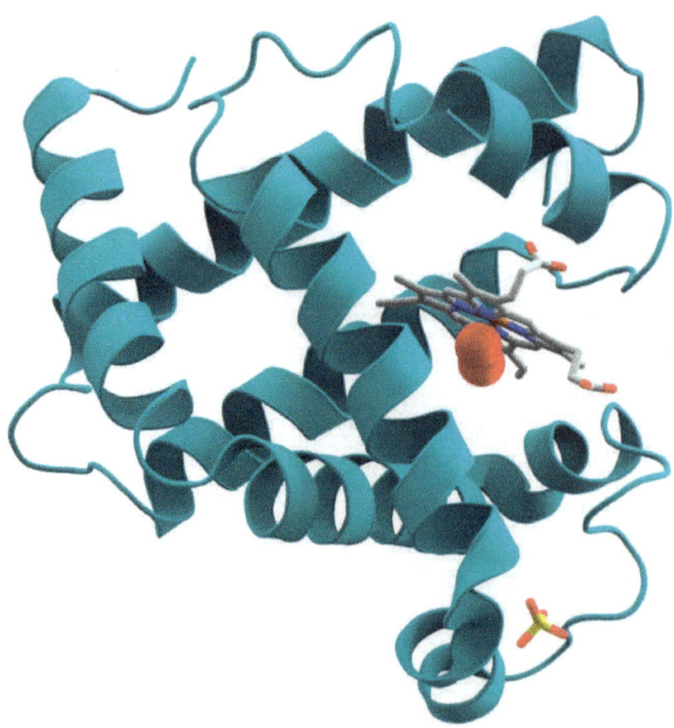

Like other biological macromolecules such as polysaccharides and nucleic acids, proteins are essential parts of organisms and participate in virtually every process within cells. Many proteins are enzymes that catalyse biochemical reactions and are vital to metabolism. Proteins also have structural or mechanical functions, such as actin and myosin in muscle and the proteins in the cytoskeleton, which form a system of scaffolding that maintains cell shape. Other proteins are important in cell signaling, immune responses, cell adhesion, and the cell cycle. In animals, proteins are needed in the diet to provide the essential amino acids that cannot be synthesized. Digestion breaks the proteins down for use in the metabolism.

Proteins may be purified from other cellular components using a variety of techniques such as ultracentrifugation, precipitation, electrophoresis, and chromatography; the advent of genetic engineering has made possible a number of methods to facilitate purification. Methods commonly used to study protein structure and function include immunohistochemistry, site-directed mutagenesis, X-ray crystallography, nuclear magnetic resonance and mass spectrometry. A representation of the 3D structure of the protein myoglobin showing turquoise α-helices. This protein was the first to have its structure solved by X-ray crystallography. Towards the right-center among the coils, a prosthetic group called a heme group (shown in gray) with a bound oxygen molecule (red).

BIOCHEMISTRY

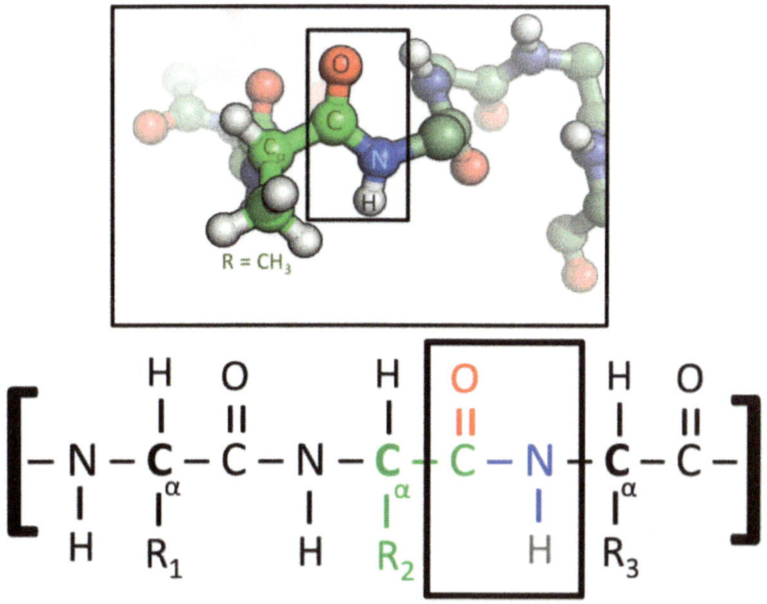

Chemical structure of the peptide bond (bottom) and the three-dimensional structure of a peptide bond between an alanine and an adjacent amino acid (top/inset)

Resonance structures of the peptide bond that links individual amino acids to form a protein polymer

Most proteins consist of linear polymers built from series of up to 20 different L-α-amino acids. All proteinogenic amino acids possess common structural features, including an α-carbon to which an amino group, a carboxyl group, and a variable side chain are bonded. Only proline differs from this basic structure as it contains an unusual ring to the N-end amine group, which forces the CO–NH amide moiety into a fixed conformation. The side chains of the standard amino acids, detailed in the list of standard amino acids, have a great variety of chemical structures and properties; it is the combined effect of all of the amino acid side chains in a protein that ultimately determines its three-dimensional structure and its chemical reactivity. The amino acids in a polypeptide chain are linked by peptide bonds. Once linked in the protein chain, an individual amino acid is called a *residue,* and the linked series of carbon, nitrogen, and oxygen atoms are known as the *main chain* or *protein backbone.*

The peptide bond has two resonance forms that contribute some double-bond character and inhibit rotation around its axis, so that the alpha carbons are roughly coplanar. The other two dihedral angles in the peptide bond determine the local shape assumed by the protein backbone. The end with a free amino group is known as the N-terminus or amino terminus, whereas the end of the protein with a free carboxyl group is known as the C-terminus or carboxy terminus (the sequence of the protein is written from N-terminus to C-terminus, from left to right).

The words *protein*, *polypeptide,* and *peptide* are a little ambiguous and can overlap in meaning. *Protein* is generally used to refer to the complete biological molecule in a stable conformation, whereas *peptide* is generally reserved for a short amino acid oligomers often lacking a stable three-dimensional structure. However, the boundary between the two is not well defined and usually lies near 20–30 residues. *Polypeptide* can refer to any single linear chain of amino acids, usually regardless of length, but often implies an absence of a defined conformation.

Abundance in Cells

It has been estimated that average-sized bacteria contain about 2 million proteins per cell (e.g. *E. coli* and *Staphylococcus aureus*). Smaller bacteria, such as *Mycoplasma* or *spirochetes* contain fewer molecules, namely on the order of 50,000 to 1 million. By contrast, eukaryotic cells are larger and thus contain much more protein. For instance, yeast cells were estimated to contain about 50 million proteins and human cells on the order of 1 to 3 billion. The concentration of individual protein copies ranges from a few molecules per cell up to 20 million. Not all genes coding proteins are expressed in most cells and their number depends on for example cell type and external stimuli. For instance, of the 20,000 or so proteins encoded by the human genome, only 6,000 are detected in lymphoblastoid cells. Moreover, the number of proteins the genome encodes correlates well with the organism complexity. Eukaryotes, bacteria, Archaea and viruses have on average 15145, 3200, 2358 and 42 proteins respectively coded in their genomes.

Synthesis

Biosynthesis

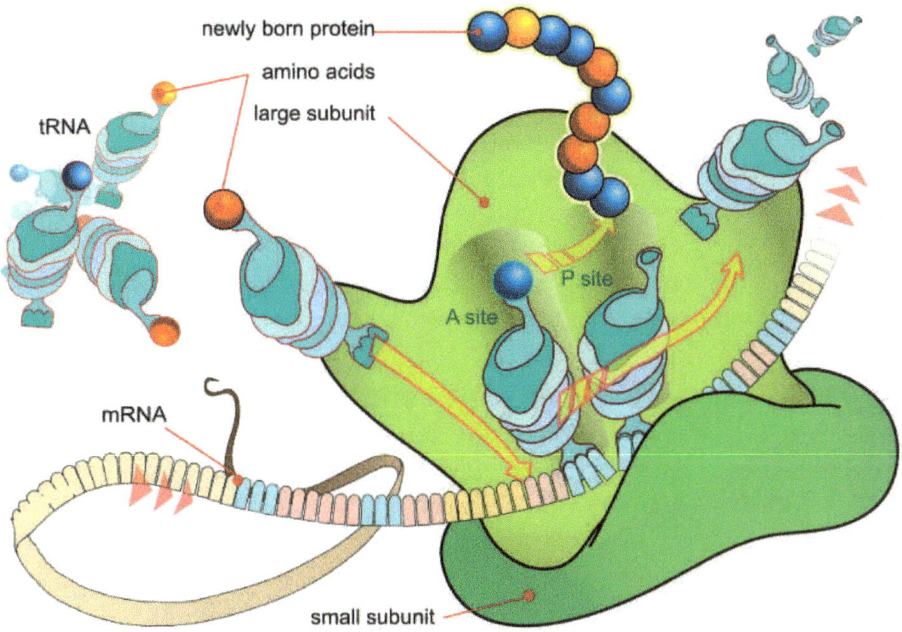

A ribosome produces a protein using mRNA as template

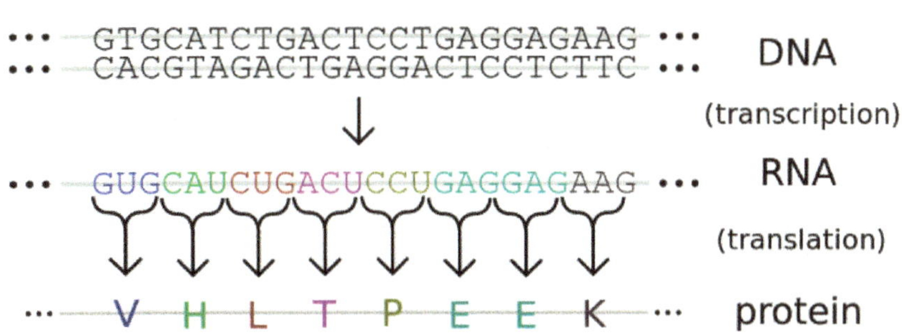

The DNA sequence of a gene encodes the amino acid sequence of a protein

Proteins are assembled from amino acids using information encoded in genes. Each protein has its own unique amino acid sequence that is specified by the nucleotide sequence of the gene encoding this protein. The genetic code is a set of three-nucleotide sets called codons and each three-nucleotide combination designates an amino acid, for example AUG (adenine-uracil-guanine) is the code for methionine. Because DNA contains four nucleotides, the total number of possible codons is 64; hence, there is some redundancy in the genetic code, with some amino acids specified by more than one codon. Genes encoded in DNA are first transcribed into pre-messenger RNA (mRNA) by proteins such as RNA polymerase. Most organisms then process the pre-mRNA (also known as a *primary transcript*) using various forms of Post-transcriptional modification to form the mature mRNA, which is then used as a template for protein synthesis by the ribosome. In prokaryotes the mRNA may either be used as soon as it is produced, or be bound by a ribosome after having moved away from the nucleoid. In contrast, eukaryotes make mRNA in the cell nucleus and then translocate it across the nuclear membrane into the cytoplasm, where protein synthesis then takes place. The rate of protein synthesis is higher in prokaryotes than eukaryotes and can reach up to 20 amino acids per second.

The process of synthesizing a protein from an mRNA template is known as translation. The mRNA is loaded onto the ribosome and is read three nucleotides at a time by matching each codon to its base pairing anticodon located on a transfer RNA molecule, which carries the amino acid corresponding to the codon it recognizes. The enzyme aminoacyl tRNA synthetase "charges" the tRNA molecules with the correct amino acids. The growing polypeptide is often termed the *nascent chain*. Proteins are always biosynthesized from N-terminus to C-terminus.

The size of a synthesized protein can be measured by the number of amino acids it contains and by its total molecular mass, which is normally reported in units of *daltons* (synonymous with atomic mass units), or the derivative unit kilodalton (kDa). The average size of protein increases from Archaea, Bacteria to Eukaryote (283, 311, 438 residues and 31, 34, 49 kDa respecitvely) due bigger number of protein domains constituting proteins in higher organisms. For instance, yeast proteins are on average 466 amino acids long and 53 kDa in mass. The largest known proteins are the titins, a component of the muscle sarcomere, with a molecular mass of almost 3,000 kDa and a total length of almost 27,000 amino acids.

Chemical Synthesis

Short proteins can also be synthesized chemically by a family of methods known as peptide synthesis, which rely on organic synthesis techniques such as chemical ligation to produce peptides in high yield. Chemical synthesis allows for the introduction of non-natural amino acids into polypeptide chains, such as attachment of fluorescent probes to amino acid side chains. These methods are useful in laboratory biochemistry and cell biology, though generally not for commercial applications. Chemical synthesis is inefficient for polypeptides longer than about 300 amino acids, and the synthesized proteins may not readily assume their native tertiary structure. Most chemical synthesis methods proceed from C-terminus to N-terminus, opposite the biological reaction.

Structure

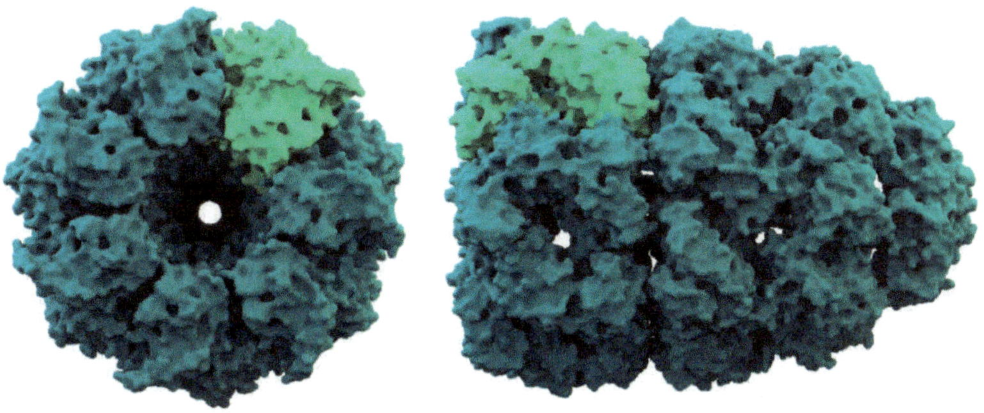

The crystal structure of the chaperonin, a huge protein complex. A single protein subunit is highlighted. Chaperonins assist protein folding.

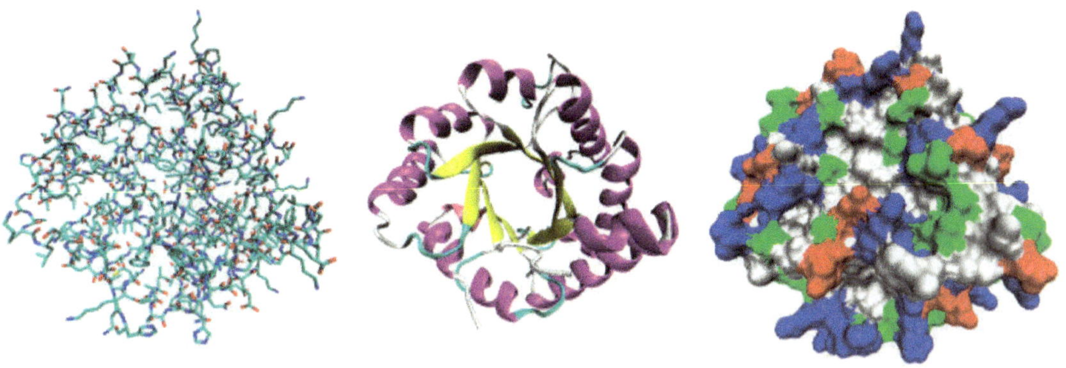

Three possible representations of the three-dimensional structure of the protein triose phosphate isomerase. Left: All-atom representation colored by atom type. Middle: Simplified representation illustrating the backbone conformation, colored by secondary structure. Right: Solvent-accessible surface representation colored by residue type (acidic residues red, basic residues blue, polar residues green, nonpolar residues white).

Most proteins fold into unique 3-dimensional structures. The shape into which a protein naturally folds is known as its native conformation.Although many proteins can fold unassisted, simply through the chemical properties of their amino acids, others require the aid of molecular chaperones to fold into their native states. Biochemists often refer to four distinct aspects of a protein's structure:

- *Primary structure*: the amino acid sequence. A protein is a polyamide.
- *Secondary structure*: regularly repeating local structures stabilized by hydrogen

bonds. The most common examples are the α-helix, β-sheet and turns. Because secondary structures are local, many regions of different secondary structure can be present in the same protein molecule.

- *Tertiary structure*: the overall shape of a single protein molecule; the spatial relationship of the secondary structures to one another. Tertiary structure is generally stabilized by nonlocal interactions, most commonly the formation of a hydrophobic core, but also through salt bridges, hydrogen bonds, disulfide bonds, and even posttranslational modifications. The term "tertiary structure" is often used as synonymous with the term *fold*. The tertiary structure is what controls the basic function of the protein.
- *Quaternary structure*: the structure formed by several protein molecules (polypeptide chains), usually called *protein subunits* in this context, which function as a single protein complex.

Proteins are not entirely rigid molecules. In addition to these levels of structure, proteins may shift between several related structures while they perform their functions. In the context of these functional rearrangements, these tertiary or quaternary structures are usually referred to as "conformations", and transitions between them are called *conformational changes*. Such changes are often induced by the binding of a substrate molecule to an enzyme's active site, or the physical region of the protein that participates in chemical catalysis. In solution proteins also undergo variation in structure through thermal vibration and the collision with other molecules.

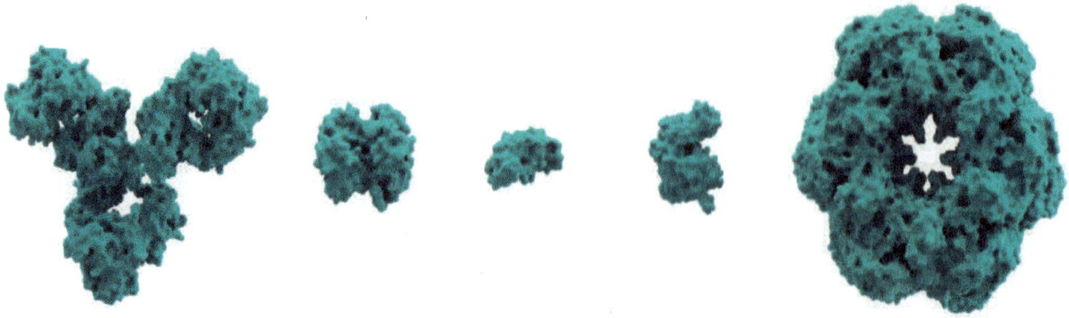

Molecular surface of several proteins showing their comparative sizes. From left to right are: immunoglobulin G (IgG, an antibody), hemoglobin, insulin (a hormone), adenylate kinase (an enzyme), and glutamine synthetase (an enzyme).

Proteins can be informally divided into three main classes, which correlate with typical tertiary structures: globular proteins, fibrous proteins, and membrane proteins. Almost all globular proteins are soluble and many are enzymes. Fibrous proteins are often structural, such as collagen, the major component of connective tissue, or keratin, the protein component of hair and nails. Membrane proteins often serve as receptors or provide channels for polar or charged molecules to pass through the cell membrane.

A special case of intramolecular hydrogen bonds within proteins, poorly shielded from water attack and hence promoting their own dehydration, are called dehydrons.

Structure Determination

Discovering the tertiary structure of a protein, or the quaternary structure of its complexes, can provide important clues about how the protein performs its function. Common experimental methods of structure determination include X-ray crystallography and NMR spectroscopy, both of which can produce information at atomic resolution. However, NMR experiments are able to provide information from which a subset of distances between pairs of atoms can be estimated, and the final possible conformations for a protein are determined by solving a distance geometry problem. Dual polarisation interferometry is a quantitative analytical method for measuring the overall protein conformation and conformational changes due to interactions or other stimulus. Circular dichroism is another laboratory technique for determining internal β-sheet / α-helical composition of proteins. Cryoelectron microscopy is used to produce lower-resolution structural information about very large protein complexes, including assembled viruses; a variant known as electron crystallography can also produce high-resolution information in some cases, especially for two-dimensional crystals of membrane proteins. Solved structures are usually deposited in the Protein Data Bank (PDB), a freely available resource from which structural data about thousands of proteins can be obtained in the form of Cartesian coordinates for each atom in the protein.

Many more gene sequences are known than protein structures. Further, the set of solved structures is biased toward proteins that can be easily subjected to the conditions required in X-ray crystallography, one of the major structure determination methods. In particular, globular proteins are comparatively easy to crystallize in preparation for X-ray crystallography. Membrane proteins, by contrast, are difficult to crystallize and are underrepresented in the PDB. Structural genomics initiatives have attempted to remedy these deficiencies by systematically solving representative structures of major fold classes. Protein structure prediction methods attempt to provide a means of generating a plausible structure for proteins whose structures have not been experimentally determined.

Cellular Functions

Proteins are the chief actors within the cell, said to be carrying out the duties specified by the information encoded in genes.With the exception of certain types of RNA, most other biological molecules are relatively inert elements upon which proteins act. Proteins make up half the dry weight of an *Escherichia coli* cell, whereas other macromolecules such as DNA and RNA make up only 3% and 20%, respectively. The set of proteins expressed in a particular cell or cell type is known as its proteome.

The chief characteristic of proteins that also allows their diverse set of functions is their ability to bind other molecules specifically and tightly. The region of the protein responsible for binding another molecule is known as the binding site and is often a depression or "pocket" on the molecular surface. This binding ability is mediated by the tertiary structure of the protein, which defines the binding site pocket, and by the chemical properties of the surrounding amino acids' side chains. Protein binding can be extraordinarily tight and specific; for example, the ribonuclease inhibitor protein binds to human angiogenin with a sub-femtomolar dissociation constant ($<10^{-15}$ M) but does not bind at all to its amphibian homolog onconase (>1 M). Extremely minor chemical changes such as the addition of a single methyl group to a binding partner can sometimes suffice to nearly eliminate binding; for example, the aminoacyl tRNA synthetase specific to the amino acid valine discriminates against the very similar side chain of the amino acid isoleucine.

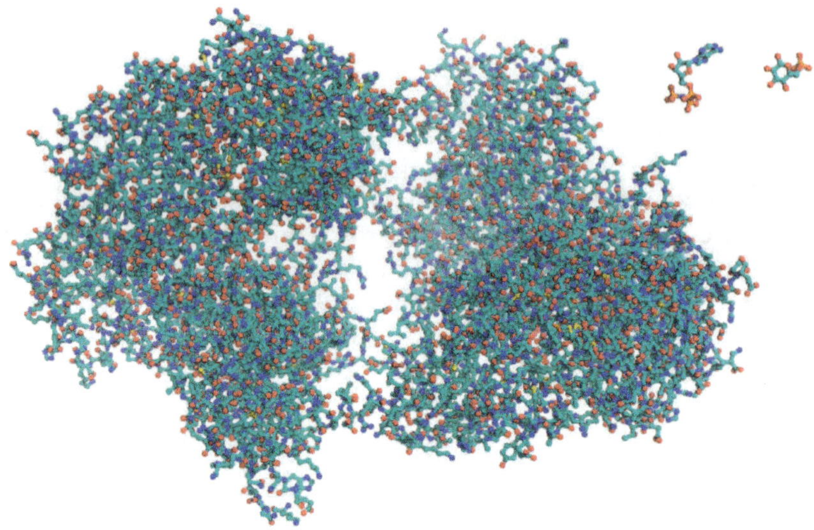

The enzyme hexokinase is shown as a conventional ball-and-stick molecular model. To scale in the top right-hand corner are two of its substrates, ATP and glucose.

Proteins can bind to other proteins as well as to small-molecule substrates. When proteins bind specifically to other copies of the same molecule, they can oligomerize to form fibrils; this process occurs often in structural proteins that consist of globular monomers that self-associate to form rigid fibers. Protein–protein interactions also regulate enzymatic activity, control progression through the cell cycle, and allow the assembly of large protein complexes that carry out many closely related reactions with a common biological function. Proteins can also bind to, or even be integrated into, cell membranes. The ability of binding partners to induce conformational changes in proteins allows the construction of enormously complex signaling networks. As interactions between proteins are reversible, and depend heavily on the availability of different groups of partner proteins to form aggregates that are capable to carry out discrete sets of function, study of the interactions between specific proteins is a key to understand important aspects of cellular function, and ultimately the properties that distinguish particular cell types.

Enzymes

The best-known role of proteins in the cell is as enzymes, which catalyse chemical reactions. Enzymes are usually highly specific and accelerate only one or a few chemical reactions. Enzymes carry out most of the reactions involved in metabolism, as well as manipulating DNA in processes such as DNA replication, DNA repair, and transcription.

Some enzymes act on other proteins to add or remove chemical groups in a process known as posttranslational modification. About 4,000 reactions are known to be catalysed by enzymes. The rate acceleration conferred by enzymatic catalysis is often enormous—as much as 10^{17}-fold increase in rate over the uncatalysed reaction in the case of orotate decarboxylase (78 million years without the enzyme, 18 milliseconds with the enzyme).

The molecules bound and acted upon by enzymes are called substrates. Although enzymes can consist of hundreds of amino acids, it is usually only a small fraction of the residues that come in contact with the substrate, and an even smaller fraction—three to four residues on average—that are directly involved in catalysis. The region of the enzyme that binds the substrate and contains the catalytic residues is known as the active site.

Dirigent proteins are members of a class of proteins that dictate the stereochemistry of a compound synthesized by other enzymes.

Cell Signaling and Ligand Binding

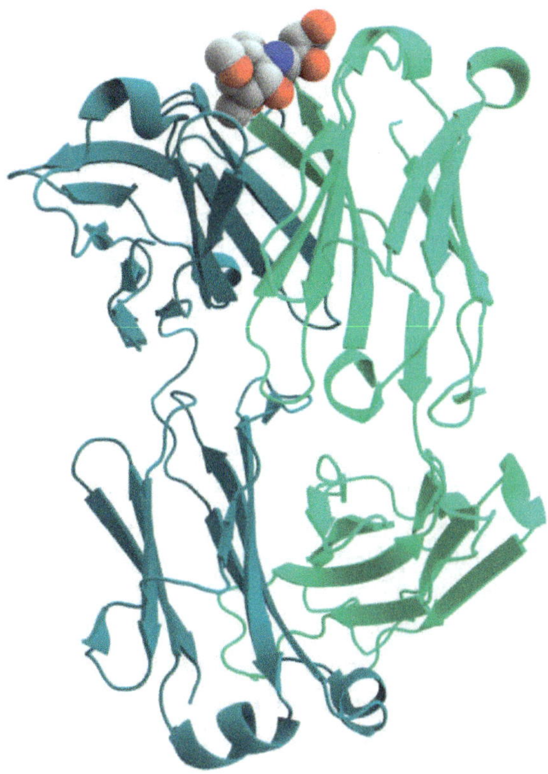

Ribbon diagram of a mouse antibody against cholera that binds a carbohydrate antigen

Many proteins are involved in the process of cell signaling and signal transduction. Some proteins, such as insulin, are extracellular proteins that transmit a signal from the cell in which they were synthesized to other cells in distant tissues. Others are membrane proteins that act as receptors whose main function is to bind a signaling molecule and induce a biochemical response in the cell. Many receptors have a binding site exposed on the cell surface and an effector domain within the cell, which may have enzymatic activity or may undergo a conformational change detected by other proteins within the cell.

Antibodies are protein components of an adaptive immune system whose main function is to bind antigens, or foreign substances in the body, and target them for destruction. Antibodies can be secreted into the extracellular environment or anchored in the membranes of specialized B cells known as plasma cells. Whereas enzymes are limited in their binding affinity for their substrates by the necessity of conducting their reaction, antibodies have no such constraints. An antibody's binding affinity to its target is extraordinarily high.

Many ligand transport proteins bind particular small biomolecules and transport them to other locations in the body of a multicellular organism. These proteins must have a high binding affinity when their ligand is present in high concentrations, but must also release the ligand when it is present at low concentrations in the target tissues. The canonical example of a ligand-binding protein is haemoglobin, which transports oxygen from the lungs to other organs and tissues in all vertebrates and has close homologs in every biological kingdom.Lectins are sugar-binding proteins which are highly specific for their sugar moieties. Lectins typically play a role in biological recognition phenomena involving cells and proteins. Receptors and hormones are highly specific binding proteins.

Transmembrane proteins can also serve as ligand transport proteins that alter the permeability of the cell membrane to small molecules and ions. The membrane alone has a hydrophobic core through which polar or charged molecules cannot diffuse. Membrane proteins contain internal channels that allow such molecules to enter and exit the cell. Many ion channel proteins are specialized to select for only a particular ion; for example, potassium and sodium channels often discriminate for only one of the two ions.

Structural Proteins

Structural proteins confer stiffness and rigidity to otherwise-fluid biological components. Most structural proteins are fibrous proteins; for example, collagen and elastin are critical components of connective tissue such as cartilage, and keratin is found in hard or filamentous structures such as hair, nails, feathers, hooves, and some animal shells. Some globular proteins can also play structural functions, for example, actin and tubulin are globular and soluble as monomers, but polymerize to form long, stiff fibers that make up the cytoskeleton, which allows the cell to maintain its shape and size.

Other proteins that serve structural functions are motor proteins such as myosin, kinesin, and dynein, which are capable of generating mechanical forces. These proteins are crucial for cellular motility of single celled organisms and the sperm of many multicellular organisms which reproduce sexually. They also generate the forces exerted by contracting muscles and play essential roles in intracellular transport.

METHODS OF STUDY

The activities and structures of proteins may be examined *in vitro, in vivo, and in silico.*

In vitro studies of purified proteins in controlled environments are useful for learning how a protein carries out its function: for example, enzyme kinetics studies explore the chemical mechanism of an enzyme's catalytic activity and its relative affinity for various possible substrate molecules. By contrast, *in vivo* experiments can provide information about the physiological role of a protein in the context of a cell or even a whole organism. *In silico* studies use computational methods to study proteins.

Protein Purification

To perform *in vitro* analysis, a protein must be purified away from other cellular components. This process usually begins with cell lysis, in which a cell's membrane is disrupted and its internal contents released into a solution known as a crude lysate. The resulting mixture can be purified using ultracentrifugation, which fractionates the various cellular components into fractions containing soluble proteins; membrane lipids and proteins; cellular organelles, and nucleic acids. Precipitation by a method known as salting out can concentrate the proteins from this lysate. Various types of chromatography are then used to isolate the protein or proteins of interest based on properties such as molecular weight, net charge and binding affinity. The level of purification can be monitored using various types of gel electrophoresis if the desired protein's molecular weight and isoelectric point are known, by spectroscopy if the protein has distinguishable spectroscopic features, or by enzyme assays if the protein has enzymatic activity. Additionally, proteins can be isolated according their charge using electrofocusing.

For natural proteins, a series of purification steps may be necessary to obtain protein sufficiently pure for laboratory applications. To simplify this process, genetic engineering is often used to add chemical features to proteins that make them easier to purify without affecting their structure or activity. Here, a "tag" consisting of a specific amino acid sequence, often a series of histidine residues (a "His-tag"), is attached to one terminus of the protein. As a result, when the lysate is passed over a chromatography column containing nickel, the histidine residues ligate the nickel and attach to the column while the untagged components of the lysate pass unimpeded. A number of different tags have been developed to help researchers purify specific proteins from complex mixtures.

Cellular Localization

The study of proteins *in vivo* is often concerned with the synthesis and localization of the protein within the cell. Although many intracellular proteins are synthesized in the cytoplasm and membrane-bound or secreted proteins in the endoplasmic reticulum, the specifics of how proteins are targeted to specific organelles or cellular structures is often unclear. A useful technique for assessing cellular localization uses genetic engineering to express in a cell a fusion protein or chimera consisting of the natural protein of interest linked to a "reporter" such as green fluorescent protein (GFP). The fused protein's position within the cell can be cleanly and efficiently visualized using microscopy, as shown in the figure opposite.

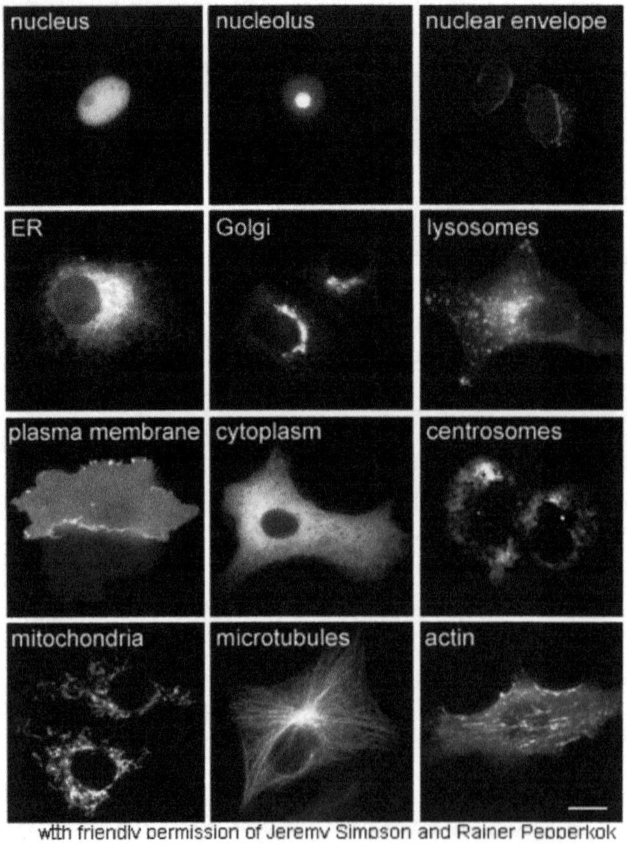

Proteins in different cellular compartments and structures tagged with green fluorescent protein (here, white)

Other methods for elucidating the cellular location of proteins requires the use of known compartmental markers for regions such as the ER, the Golgi, lysosomes or vacuoles, mitochondria, chloroplasts, plasma membrane, etc. With the use of fluorescently tagged versions of these markers or of antibodies to known markers, it becomes much simpler to identify the localization of a protein of interest. For example, indirect immunofluorescence will allow for fluorescence colocalization and demonstration of location. Fluorescent dyes are used to label cellular compartments for a similar purpose.

Other possibilities exist, as well. For example, immunohistochemistry usually utilizes an antibody to one or more proteins of interest that are conjugated to enzymes yielding either luminescent or chromogenic signals that can be compared between samples, allowing for localization information. Another applicable technique is cofractionation in sucrose (or other material) gradients using isopycnic centrifugation. While this technique does not prove colocalization of a compartment of known density and the protein of interest, it does increase the likelihood, and is more amenable to large-scale studies.

Finally, the gold-standard method of cellular localization is immunoelectron microscopy. This technique also uses an antibody to the protein of interest, along with classical electron microscopy techniques. The sample is prepared for normal electron microscopic examination, and then treated with an antibody to the protein of interest that is conjugated to an extremely electro-dense material, usually gold. This allows for the localization of both ultrastructural details as well as the protein of interest.

Through another genetic engineering application known as site-directed mutagenesis, researchers can alter the protein sequence and hence its structure, cellular localization, and susceptibility to regulation. This technique even allows the incorporation of unnatural amino acids into proteins, using modified tRNAs, and may allow the rational design of new proteins with novel properties.

Proteomics

The total complement of proteins present at a time in a cell or cell type is known as its proteome, and the study of such large-scale data sets defines the field of proteomics, named by analogy to the related field of genomics. Key experimental techniques in proteomics include 2D electrophoresis, which allows the separation of a large number of proteins, mass spectrometry, which allows rapid high-throughput identification of proteins and sequencing of peptides (most often after in-gel digestion), protein microarrays, which allow the detection of the relative levels of a large number of proteins present in a cell, and two-hybrid screening, which allows the systematic exploration of protein–protein interactions. The total complement of biologically possible such interactions is known as the interactome. A systematic attempt to determine the structures of proteins representing every possible fold is known as structural genomics.

Bioinformatics

A vast array of computational methods have been developed to analyze the structure, function, and evolution of proteins.

The development of such tools has been driven by the large amount of genomic and proteomic data available for a variety of organisms, including the human genome. It is simply impossible to study all proteins experimentally, hence only a few are subjected to laboratory experiments while computational tools are used to extrapolate to similar proteins. Such homologous proteins can be efficiently identified in distantly related organisms by sequence alignment. Genome and gene sequences can be searched by a variety of tools for certain properties. Sequence profiling tools can find restriction enzyme sites, open reading frames in nucleotide sequences, and predict secondary structures. Phylogenetic trees can be constructed and evolutionary hypotheses developed using special software like ClustalW regarding the ancestry of modern organisms and the genes they express. The field of bioinformatics is now indispensable for the analysis of genes and proteins.

Structure Prediction and Simulation

Complementary to the field of structural genomics, protein structure prediction seeks to develop efficient ways to provide plausible models for proteins whose structures have not yet been determined experimentally. The most successful type of structure prediction, known as homology modeling, relies on the existence of a "template" structure with sequence similarity to the protein being modeled; structural genomics' goal is to provide sufficient representation in solved structures to model most of those that remain. Although producing accurate models remains a challenge when only distantly related template structures are available, it has been suggested that sequence alignment is the bottleneck in this process, as quite accurate models can be produced if a "perfect" sequence alignment is known. Many structure prediction methods have served to inform the emerging field of protein engineering, in which novel protein folds have already been designed. A more complex computational problem is the prediction of intermolecular interactions, such as in molecular docking and protein–protein interaction prediction.

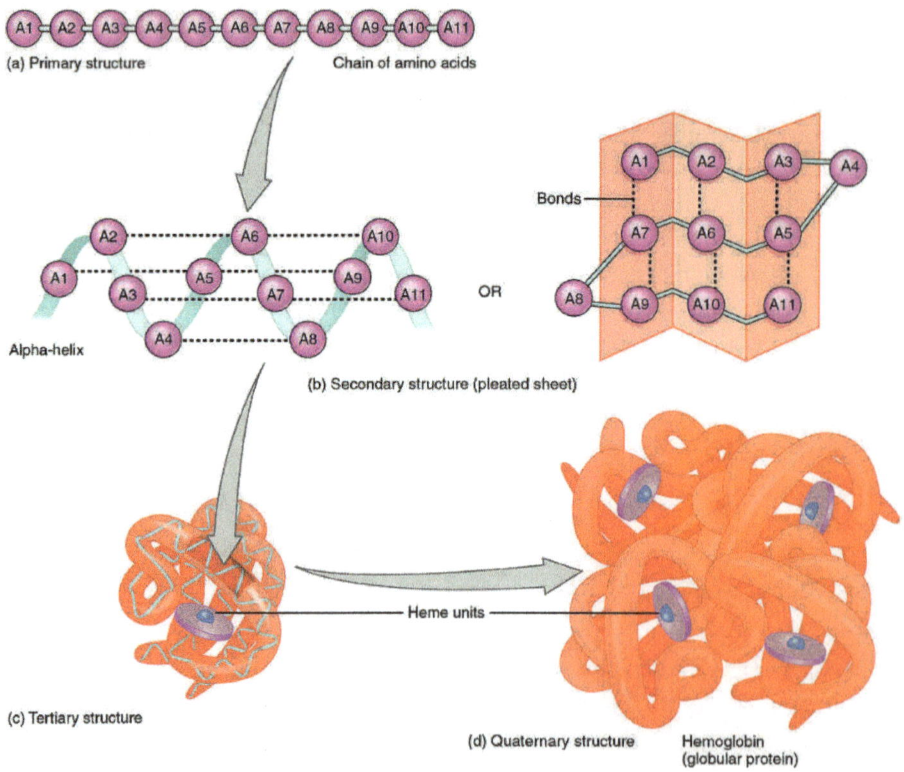

Constituent amino-acids can be analyzed to predict secondary, tertiary and quaternary protein structure, in this case hemoglobin containing heme units

The processes of protein folding and binding can be simulated using such technique as molecular mechanics, in particular, molecular dynamics and Monte Carlo, which increasingly take advantage of parallel and distributed computing (Folding@home project; molecular modeling on GPU). The folding of small α-helical protein domains such as the villin headpiece and the HIV accessory protein have been successfully simulated *in silico*, and hybrid methods that combine standard molecular dynamics with quantum mechanics calculations have allowed exploration of the electronic states of rhodopsins.

Protein disorder and Unstructure Prediction

Many proteins (in Eucaryota ~33%) contain large unstructured but biologically functional segments and can be classified as intrinsically disordered proteins. Predicting and analysing protein disorder is, therefore, an important part of protein structure characterisation.

Nutrition

Most microorganisms and plants can biosynthesize all 20 standard amino acids, while animals (including humans) must obtain some of the amino acids from the diet. The amino acids that an organism cannot synthesize on its own are referred to as essential amino acids. Key enzymes that synthesize certain amino acids are not present in animals — such as aspartokinase, which catalyses the first step in the synthesis of lysine, methionine, and threonine from aspartate. If amino acids are present in the environment, microorganisms can conserve energy by taking up the amino acids from their surroundings and downregulating their biosynthetic pathways.

In animals, amino acids are obtained through the consumption of foods containing protein. Ingested proteins are then broken down into amino acids through digestion, which typically involves denaturation of the protein through exposure to acid and hydrolysis by enzymes called proteases. Some ingested amino acids are used for protein biosynthesis, while others are converted to glucose through gluconeogenesis, or fed into the citric acid cycle. This use of protein as a fuel is particularly important under starvation conditions as it allows the body's own proteins to be used to support life, particularly those found in muscle. Amino acids are also an important dietary source of nitrogen.

HISTORY AND ETYMOLOGY

Proteins were recognized as a distinct class of biological molecules in the eighteenth century by Antoine Fourcroy and others, distinguished by the molecules' ability to coagulate or flocculate under treatments with heat or acid. Noted examples at the time included albumin from egg whites, blood serum albumin, fibrin, and wheat gluten.

Proteins were first described by the Dutch chemist Gerardus Johannes Mulder and named by the Swedish chemist Jöns Jacob Berzelius in 1838. Mulder carried out elemental analysis of common proteins and found that nearly all proteins had the same empirical formula, $C_{400}H_{620}N_{100}O_{120}P_1S_1$. He came to the erroneous conclusion that they might be composed of a single type of (very large) molecule. The term "protein" to describe these molecules was proposed by Mulder's associate Berzelius; protein is derived from the Greek word πρώτειος (*proteios*), meaning "primary", "in the lead", or "standing in front", + *-in*. Mulder went on to identify the products of protein degradation such as the amino acid leucine for which he found a (nearly correct) molecular weight of 131 Da.

Early nutritional scientists such as the German Carl von Voit believed that protein was the most important nutrient for maintaining the structure of the body, because it was generally believed that "flesh makes flesh." Karl Heinrich Ritthausen extended known protein forms with the identification of glutamic acid. At the Connecticut Agricultural Experiment Station a detailed review of the vegetable proteins was compiled by Thomas Burr Osborne. Working with Lafayette Mendel and applying Liebig's law of the minimum in feeding laboratory rats, the nutritionally essential amino acids were established. The work was continued and communicated by William Cumming Rose. The understanding of proteins as polypeptides came through the work of Franz Hofmeister and Hermann Emil Fischer. The central role of proteins as enzymes in living organisms was not fully appreciated until 1926, when James B. Sumner showed that the enzyme urease was in fact a protein.

The difficulty in purifying proteins in large quantities made them very difficult for early protein biochemists to study. Hence, early studies focused on proteins that could be purified in large quantities, e.g., those of blood, egg white, various toxins, and digestive/ metabolic enzymes obtained from slaughterhouses. In the 1950s, the Armour Hot Dog Co. purified 1 kg of pure bovine pancreatic ribonuclease A and made it freely available to scientists; this gesture helped ribonuclease A become a major target for biochemical study for the following decades.

John Kendrew with model of myoglobin in progress

Linus Pauling is credited with the successful prediction of regular protein secondary structures based on hydrogen bonding, an idea first put forth by William Astbury in 1933. Later work by Walter Kauzmann on denaturation, based partly on previous studies by Kaj Linderstrøm-Lang,contributed an understanding of protein folding and structure mediated by hydrophobic interactions.

The first protein to be sequenced was insulin, by Frederick Sanger, in 1949. Sanger correctly determined the amino acid sequence of insulin, thus conclusively demonstrating that proteins consisted of linear polymers of amino acids rather than branched chains, colloids, or cyclols. He won the Nobel Prize for this achievement in 1958.

The first protein structures to be solved were hemoglobin and myoglobin, by Max Perutz and Sir John Cowdery Kendrew, respectively, in 1958. As of 2016, the Protein Data Bank has over 115,000 atomic-resolution structures of proteins. In more recent times, cryo-electron microscopy of large macromolecular assemblies and computational protein structure prediction of small protein domains are two methods approaching atomic resolution.

REFERENCES

1. Nelson DL, Cox MM (2005). Lehninger's Principles of Biochemistry (4th ed.). New York, New York: W. H. Freeman and Company.
2. Gutteridge A, Thornton JM (2005). "Understanding nature's catalytic toolkit". Trends in Biochemical Sciences. 30 (11): 622–29. doi:10.1016/j.tibs.2005.09.006. PMID 16214343.
3. Murray et al., p. 19.
4. Murray et al., p. 31.
5. Lodish H, Berk A, Matsudaira P, Kaiser CA, Krieger M, Scott MP, Zipurksy SL, Darnell J (2004). Molecular Cell Biology (5th ed.). New York, New York: WH Freeman and Company.
6. Milo, Ron (2013-12-01). "What is the total number of protein molecules per cell volume? A call to rethink some published values". BioEssays. 35 (12): 1050–1055. doi:10.1002/bies.201300066. ISSN 1521-1878. PMC 3910158 . PMID 24114984.
7. Beck, M.; Schmidt, A.; Malmstroem, J.; Claassen, M.; Ori, A.; Szymborska, A.; Herzog, F.; Rinner, O.; Ellenberg, J. (2011). "The quantitative proteome of a human cell line.". Mol Syst Biol. 7: 549. doi:10.1038/msb.2011.82. PMID 22068332.
8. Wu, Linfeng; Candille, Sophie I.; Choi, Yoonha; Xie, Dan; Jiang, Lihua; Li-Pook-Than, Jennifer; Tang, Hua; Snyder, Michael. "Variation and genetic control of protein abundance in humans". Nature. 499 (7456): 79–82. doi:10.1038/nature12223. PMC 3789121 . PMID 23676674.
9. Kozlowski, Lukasz P. (2016). "Proteome-pI: proteome isoelectric point database". Nucleic Acids Research: gkw978. doi:10.1093/nar/gkw978. ISSN 0305-1048.
10. van Holde and Mathews, pp. 1002–42.
11. Dobson CM (2000). "The nature and significance of protein folding". In Pain RH. Mechanisms of Protein Folding. Oxford, Oxfordshire: Oxford University Press. pp. 1–28. ISBN 0-19-963789-X.
12. Fulton A, Isaacs W (1991). "Titin, a huge, elastic sarcomeric protein with a probable role in morphogenesis". BioEssays. 13 (4): 157–61. doi:10.1002/bies.950130403. PMID 1859393.
13. Bruckdorfer T, Marder O, Albericio F (2004). "From production of peptides in milligram amounts for research to multi-tons quantities for drugs of the future". Current Pharmaceutical Biotechnology. 5 (1): 29–43. doi:10.2174/1389201043489620. PMID 14965208.
14. Schwarzer D, Cole P (2005). "Protein semisynthesis and expressed protein ligation: chasing a protein's tail". Current Opinion in Chemical Biology. 9 (6): 561–69. doi:10.1016/j.cbpa.2005.09.018. PMID 16226484.
15. Kent SB (2009). "Total chemical synthesis of proteins". Chemical Society Reviews. 38 (2): 338–51. doi:10.1039/b700141j. PMID 19169452.

16. Murray et al., p. 36.
17. Murray et al., p. 37.
18. Murray et al., pp. 30–34.
19. van Holde and Mathews, pp. 368–75.
20. van Holde and Mathews, pp. 165–85.
21. Fernández A, Scott R (2003). "Dehydron: a structurally encoded signal for protein interaction". Biophysical Journal. 85 (3): 1914–28. Bibcode:2003BpJ....85.1914F. doi:10.1016/S0006-3495(03)74619-0. PMC 1303363 . PMID 12944304.
22. Branden and Tooze, pp. 340–41.
23. Gonen T, Cheng Y, Sliz P, Hiroaki Y, Fujiyoshi Y, Harrison SC, Walz T (2005). "Lipid-protein interactions in double-layered two-dimensional AQP0 crystals". Nature. 438 (7068): 633–38. Bibcode:2005Natur.438..633G. doi:10.1038/nature04321. PMC 1350984 . PMID 16319884.
24. Standley DM, Kinjo AR, Kinoshita K, Nakamura H (2008). "Protein structure databases with new web services for structural biology and biomedical research". Briefings in Bioinformatics. 9 (4): 276–85. doi:10.1093/bib/bbn015. PMID 18430752.
25. Walian P, Cross TA, Jap BK (2004). "Structural genomics of membrane proteins". Genome Biology. 5 (4): 215. doi:10.1186/gb-2004-5-4-215. PMC 395774 . PMID 15059248.
26. Sleator RD. (2012). "Prediction of protein functions". Methods in Molecular Biology. Methods in Molecular Biology. 815: 15–24. doi:10.1007/978-1-61779-424-7_2. ISBN 978-1-61779-423-0. PMID 22130980.
27. Voet D, Voet JG. (2004). Biochemistry Vol 1 3rd ed. Wiley: Hoboken, NJ.
28. Sankaranarayanan R, Moras D (2001). "The fidelity of the translation of the genetic code". Acta Biochimica Polonica. 48 (2): 323–35. PMID 11732604.
29. van Holde and Mathews, pp. 830–49.
30. Copland JA, Sheffield-Moore M, Koldzic-Zivanovic N, Gentry S, Lamprou G, Tzortzatou-Stathopoulou F, Zoumpourlis V, Urban RJ, Vlahopoulos SA (2009). "Sex steroid receptors in skeletal differentiation and epithelial neoplasia: is tissue-specific intervention possible?". BioEssays. 31 (6): 629–41. doi:10.1002/bies.200800138. PMID 19382224.
31. Samarin S, Nusrat A (2009). "Regulation of epithelial apical junctional complex by Rho family GTPases". Frontiers in Bioscience. 14 (14): 1129–42. doi:10.2741/3298. PMID 19273120.
32. Bairoch A (2000). "The ENZYME database in 2000"(PDF). Nucleic Acids Research. 28 (1): 304–305. doi:10.1093/nar/28.1.304. PMC 102465 . PMID 10592255. Archived from the original (PDF) on June 1, 2011.
33. Radzicka A, Wolfenden R (1995). "A proficient enzyme". Science. 267 (5194): 90–93. Bibcode:1995Sci...267...90R. doi:10.1126/science.7809611. PMID 7809611.
34. EBI External Services (2010-01-20). "The Catalytic Site Atlas at The European

Bioinformatics Institute". Ebi.ac.uk. Retrieved 2011-01-16.

35. Pickel B, Schaller A (2013). "Dirigent proteins: molecular characteristics and potential biotechnological applications". Applied Microbiology and Biotechnology. 97 (19): 8427–8438. doi:10.1007/s00253-013-5167-4. PMID 23989917.
36. Branden and Tooze, pp. 251–81.
37. van Holde and Mathews, pp. 247–50.
38. van Holde and Mathews, pp. 220–29.
39. Rüdiger H, Siebert HC, Solís D, Jiménez-Barbero J, Romero A, von der Lieth CW, Diaz-Mariño T, Gabius HJ (2000). "Medicinal chemistry based on the sugar code: fundamentals of lectinology and experimental strategies with lectins as targets". Current Medicinal Chemistry. 7 (4): 389–416. doi:10.2174/0929867003375164. PMID 10702616.
40. Branden and Tooze, pp. 232–34.
41. van Holde and Mathews, pp. 178–81.
42. van Holde and Mathews, pp. 258–64; 272.
43. Murray et al., pp. 21–24.
44. Hey J, Posch A, Cohen A, Liu N, Harbers A (2008). "Fractionation of complex protein mixtures by liquid-phase isoelectric focusing". Methods in Molecular Biology. Methods in Molecular Biology™. 424: 225–39. doi:10.1007/978-1-60327-064-9_19. ISBN 978-1-58829-722-8. PMID 18369866.
45. Terpe K (2003). "Overview of tag protein fusions: from molecular and biochemical fundamentals to commercial systems". Applied Microbiology and Biotechnology. 60 (5): 523–33. doi:10.1007/s00253-002-1158-6. PMID 12536251.
46. Stepanenko OV, Verkhusha VV, Kuznetsova IM, Uversky VN, Turoverov KK (2008). "Fluorescent proteins as biomarkers and biosensors: throwing color lights on molecular and cellular processes". Current Protein & Peptide Science. 9 (4): 338–69. doi:10.2174/138920308785132668. PMC 2904242 . PMID 18691124.
47. Yuste R (2005). "Fluorescence microscopy today". Nature Methods. 2 (12): 902–904. doi:10.1038/nmeth1205-902. PMID 16299474.
48. Margolin W (2000). "Green fluorescent protein as a reporter for macromolecular localization in bacterial cells". Methods (San Diego, Calif.). 20 (1): 62–72. doi:10.1006/meth.1999.0906. PMID 10610805.
49. Walker JH, Wilson K (2000). Principles and Techniques of Practical Biochemistry. Cambridge, UK: Cambridge University Press. pp. 287–89. ISBN 0-521-65873-X.
50. Mayhew TM, Lucocq JM (2008). "Developments in cell biology for quantitative immunoelectron microscopy based on thin sections: a review". Histochemistry and Cell Biology. 130 (2): 299–313. doi:10.1007/s00418-008-0451-6. PMC 2491712 . PMID 18553098.
51. Hohsaka T, Sisido M (2002). "Incorporation of non-natural amino acids into proteins". Current Opinion in Chemical Biology. 6 (6): 809–15. doi:10.1016/S1367-5931(02)00376-9. PMID 12470735.

52. Cedrone F, Ménez A, Quéméneur E (2000). "Tailoring new enzyme functions by rational redesign". Current Opinion in Structural Biology. 10 (4): 405–10. doi:10.1016/S0959-440X(00)00106-8. PMID 10981626.

53. Görg A, Weiss W, Dunn MJ (2004). "Current two-dimensional electrophoresis technology for proteomics". Proteomics. 4 (12): 3665–85. doi:10.1002/pmic.200401031. PMID 15543535.

54. Conrotto P, Souchelnytskyi S (2008). "Proteomic approaches in biological and medical sciences: principles and applications". Experimental Oncology. 30 (3): 171–80. PMID 18806738.

55. Joos T, Bachmann J (2009). "Protein microarrays: potentials and limitations". Frontiers in Bioscience. 14 (14): 4376–85. doi:10.2741/3534. PMID 19273356.

56. Koegl M, Uetz P (2007). "Improving yeast two-hybrid screening systems". Briefings in Functional Genomics & Proteomics. 6 (4): 302–12. doi:10.1093/bfgp/elm035. PMID 18218650.

57. Plewczyński D, Ginalski K (2009). "The interactome: predicting the protein–protein interactions in cells". Cellular & Molecular Biology Letters. 14 (1): 1–22. doi:10.2478/s11658-008-0024-7. PMID 18839074.

58. Zhang C, Kim SH (2003). "Overview of structural genomics: from structure to function". Current Opinion in Chemical Biology. 7 (1): 28–32. doi:10.1016/S1367-5931(02)00015-7. PMID 12547423.

59. Zhang Y (2008). "Progress and challenges in protein structure prediction". Current Opinion in Structural Biology. 18 (3): 342–48. doi:10.1016/j.sbi.2008.02.004. PMC 2680823 . PMID 18436442.

60. Xiang Z (2006). "Advances in homology protein structure modeling". Current Protein and Peptide Science. 7 (3): 217–27. doi:10.2174/138920306777452312. PMC 1839925 . PMID 16787261.

61. Zhang Y, Skolnick J (2005). "The protein structure prediction problem could be solved using the current PDB library". Proceedings of the National Academy of Sciences of the United States of America. 102 (4): 1029–34. Bibcode:2005PNAS..102.1029Z. doi:10.1073/pnas.0407152101. PMC 545829 . PMID 15653774.

62. Kuhlman B, Dantas G, Ireton GC, Varani G, Stoddard BL, Baker D (2003). "Design of a novel globular protein fold with atomic-level accuracy". Science. 302 (5649): 1364–68. Bibcode:2003Sci...302.1364K. doi:10.1126/science.1089427. PMID 14631033.

63. Ritchie DW (2008). "Recent progress and future directions in protein–protein docking". Current Protein and Peptide Science. 9 (1): 1–15. doi:10.2174/138920308783565741. PMID 18336319.

64. Scheraga HA, Khalili M, Liwo A (2007). "Protein-folding dynamics: overview of molecular simulation techniques". Annual Review of Physical Chemistry. 58: 57–83. Bibcode:2007ARPC...58...57S. doi:10.1146/annurev.physchem.58.032806.104614. PMID 17034338.

65. Zagrovic B, Snow CD, Shirts MR, Pande VS (2002). "Simulation of folding of a small alpha-helical protein in atomistic detail using worldwide-distributed computing". Journal of Molecular Biology. 323 (5): 927–37. doi:10.1016/S0022-2836(02)00997-X. PMID 12417204.

66. Herges T, Wenzel W (2005). "In silico folding of a three helix protein and characterization of its free-energy landscape in an all-atom force field". Physical Review Letters. 94 (1): 018101. Bibcode:2005PhRvL..94a8101H. doi:10.1103/PhysRevLett.94.018101. PMID 15698135.

67. Hoffmann M, Wanko M, Strodel P, König PH, Frauenheim T, Schulten K, Thiel W, Tajkhorshid E, Elstner M (2006). "Color tuning in rhodopsins: the mechanism for the spectral shift between bacteriorhodopsin and sensory rhodopsin II". Journal of the American Chemical Society. 128 (33): 10808–18. doi:10.1021/ja062082i. PMID 16910676.

68. Ward, J. J.; Sodhi, J. S.; McGuffin, L. J.; Buxton, B. F.; Jones, D. T. (2004). "Prediction and functional analysis of native disorder in proteins from the three kingdoms of life". Journal of Molecular Biology. 337 (3): 635–45. doi:10.1016/j.jmb.2004.02.002. PMID 15019783.

69. Peter Tompa; Alan Fersht (18 November 2009). Structure and Function of Intrinsically Disordered Proteins. CRC Press. ISBN 978-1-4200-7893-0.

70. Brosnan J (June 2003). "Interorgan amino acid transport and its regulation". Journal of Nutrition. 133 (6 Suppl 1): 2068S–72S. PMID 12771367.

71. Thomas Burr Osborne (1909): The Vegetable Proteins, History pp 1 to 6, from archive.org

72. Bulletin des Sciences Physiques et Naturelles en Néerlande (1838). pg 104. SUR LA COMPOSITION DE QUELQUES SUBSTANCES ANIMALES

73. Hartley Harold (1951). "Origin of the Word 'Protein.'". Nature. 168 (4267): 244–244. doi:10.1038/168244a0.

74. Perrett D (2007). "From 'protein' to the beginnings of clinical proteomics". Proteomics: Clinical Applications. 1 (8): 720–38. doi:10.1002/prca.200700525. PMID 21136729.

75. New Oxford Dictionary of English

76. Reynolds JA, Tanford C (2003). Nature's Robots: A History of Proteins (Oxford Paperbacks). New York, New York: Oxford University Press. p. 15. ISBN 0-19-860694-X.

77. Bischoff TLW, Voit, C (1860). Die Gesetze der Ernaehrung des Pflanzenfressers durch neue Untersuchungen festgestellt (in German). Leipzig, Heidelberg.

78. Sumner JB (1926). "The isolation and crystallization of the enzyme urease. Preliminary paper" (PDF). Journal of Biological Chemistry. 69 (2): 435–41.

79. Pauling L, Corey RB, Branson HR (1951). "The structure of proteins: two hydrogen-bonded helical configurations of the polypeptide chain" (PDF). Proceedings of the National Academy of Sciences of the United States of America. 37 (5): 235–40. Bibcode:1951PNAS...37..235P. doi:10.1073/pnas.37.5.235. PMC

1063348 . PMID 14834145.

80. Kauzmann W (1956). “Structural factors in protein denaturation”. Journal of Cellular Physiology. Supplement. 47 (Suppl 1): 113–31. doi:10.1002/jcp.1030470410. PMID 13332017.
81. Kauzmann W (1959). “Some factors in the interpretation of protein denaturation”. Advances in Protein Chemistry. Advances in Protein Chemistry. 14: 1–63. doi:10.1016/S0065-3233(08)60608-7. ISBN 978-0-12-034214-3. PMID 14404936.
82. Kalman SM, Linderstrom-Lang K, Ottesen M, Richards FM (1955). “Degradation of ribonuclease by subtilisin”. Biochimica et Biophysica Acta. 16 (2): 297–99. doi:10.1016/0006-3002(55)90224-9. PMID 14363272.
83. Sanger F (1949). “The terminal peptides of insulin”. Biochemical Journal. 45 (5): 563–74. PMC 1275055 . PMID 15396627.
84. Sanger F. (1958), Nobel lecture: The chemistry of insulin(PDF), Nobelprize.org
85. Muirhead H, Perutz M (1963). “Structure of hemoglobin. A three-dimensional fourier synthesis of reduced human hemoglobin at 5.5 Å resolution”. Nature. 199 (4894): 633–38. Bibcode:1963Natur.199..633M. doi:10.1038/199633a0. PMID 14074546.
86. Kendrew J, Bodo G, Dintzis H, Parrish R, Wyckoff H, Phillips D (1958). “A three-dimensional model of the myoglobin molecule obtained by X-ray analysis”. Nature. 181 (4610): 662–66. Bibcode:1958Natur.181..662K. doi:10.1038/181662a0. PMID 13517261.
87. “RCSB Protein Data Bank”. Retrieved 2014-04-05.
88. Zhou ZH (2008). “Towards atomic resolution structural determination by single-particle cryo-electron microscopy”. Current Opinion in Structural Biology. 18 (2): 218–28. doi:10.1016/j.sbi.2008.03.004. PMC 2714865 . PMID 18403197.
89. Keskin O, Tuncbag N, Gursoy A (2008). “Characterization and prediction of protein interfaces to infer protein-protein interaction networks”. Current Pharmaceutical Biotechnology. 9 (2): 67–76. doi:10.2174/138920108783955191. PMID 18393863.

CHAPTER 2

Nucleic Acid Biomolecules

Chapter Contents

- **Introduction**
- **History of Nucleic Acids**
- **Occurrence and Nomenclature**
- **Nucleic Acid Analogue**
- **Backbone Analogues**

Every living organism needs to carry with it instructions to build a new version of itself, and to make products to keep it alive. These instructions are nucleic acids. All organisms with more than one cell use a nucleic acid called deoxyribonucleic acid (DNA), and less complicated ones, such as viruses, use ribonucleic acid (RNA.) Each of these nucleic acids is a string of many individual molecules, and when the life form reads these molecules in sequence, it identifies what product to make.

INTRODUCTION

Nucleic acids are biopolymers, or large biomolecules, essential for all known forms of life. Nucleic acids, which include DNA (deoxyribonucleic acid) and RNA (ribonucleic acid), are made from monomers known as nucleotides. Each nucleotide has three components: a 5-carbon sugar, a phosphate group, and a nitrogenous base. If the sugar is deoxyribose, the polymer is DNA. If the sugar is ribose, the polymer is RNA. When all three components are combined, they form a nucleotide. Nucleotides are also known as phosphate nucleotides.

Nucleic acids are among the most important biological macromolecules (others being amino acids-proteins, sugars-carbohydrates, and lipids-fats). They are found in abundance in all living things, where they function in encoding, transmitting and expressing genetic information. In other words, information is conveyed through the nucleic acid sequence, or the order of nucleotides within a DNA or RNA molecule. Strings of nucleotides strung together in a specific sequence are the mechanism for storing and transmitting hereditary, or genetic information via protein synthesis.

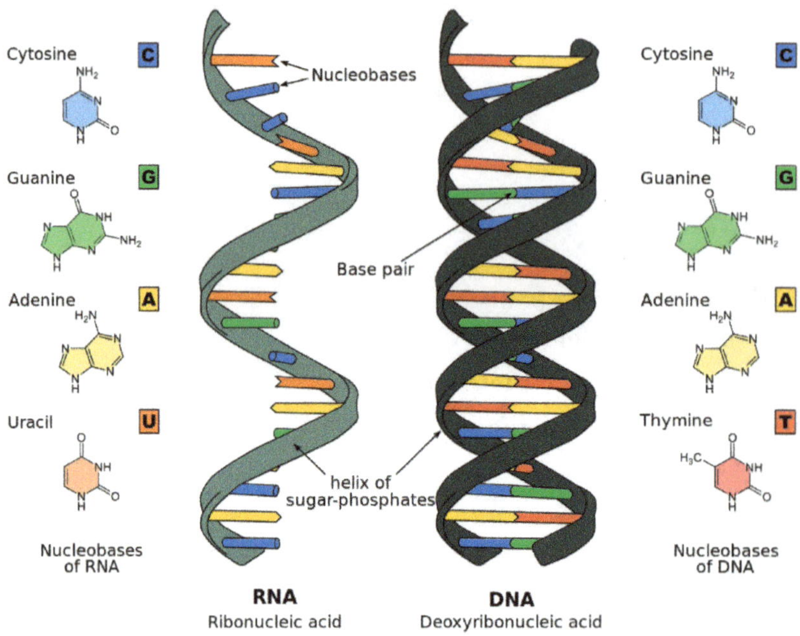

A comparison of the two principal nucleic acids: RNA (left) and DNA (right), showing the helices and nucleobases each employs.

The Swiss scientist Friedrich Miescher discovered nucleic acids (DNA) in 1869. Later, he raised the idea that they could be involved in heredity.

HISTORY OF NUCLEIC ACIDS

- Nucleic acids were discovered by Friedrich Miescher in 1869.
- In **1889** Richard Altmann discovered that nuclein has acidic properties, and it became called nucleic acid
- In **1938** Astbury and Bell published the first X-ray diffraction pattern of DNA.
- In **1953** Watson and Crick determined the structure of DNA.

Experimental studies of nucleic acids constitute a major part of modern biological and medical research, and form a foundation for genome and forensic science, and the biotechnology and pharmaceutical industries.

OCCURRENCE AND NOMENCLATURE

The term *nucleic acid* is the overall name for DNA and RNA, members of a family of biopolymers, and is synonymous with *polynucleotide*. Nucleic acids were named for their initial discovery within the nucleus, and for the presence of phosphate groups (related to phosphoric acid). Although first discovered within the nucleus of eukaryotic cells, nucleic acids are now known to be found in all life forms including within bacteria, archaea, mitochondria, chloroplasts, viruses, and viroids. (note: there is debate as to whether viruses are living or non-living). All living cells contain both DNA and RNA (except some cells such as mature red blood cells), while viruses contain either DNA or RNA, but usually not both. The basic component of biological nucleic acids is the nucleotide, each of which contains a pentose sugar (ribose or deoxyribose), a phosphate group, and a nucleobase. Nucleic acids are also generated within the laboratory, through the use of enzymes (DNA and RNA polymerases) and by solid-phase chemical synthesis. The chemical methods also enable the generation of altered nucleic acids that are not found in nature, for example peptide nucleic acids.

Molecular Composition and Size

Nucleic acids are generally very large molecules. Indeed, DNA molecules are probably the largest individual molecules known. Well-studied biological nucleic acid molecules range in size from 21 nucleotides (small interfering RNA) to large chromosomes (human chromosome 1 is a single molecule that contains 247 million base pairs).

In most cases, naturally occurring DNA molecules are double-stranded and RNA molecules are single-stranded. There are numerous exceptions, however—some viruses have genomes made of double-stranded RNA and other viruses have single-stranded DNA genomes, and, in some circumstances, nucleic acid structures with three or four strands can form.

Nucleic acids are linear polymers (chains) of nucleotides. Each nucleotide consists of three components: a purine or pyrimidinenucleobase (sometimes termed *nitrogenous base* or simply *base*), a pentose sugar, and a phosphate group. The substructure consisting of a nucleobase plus sugar is termed a nucleoside. Nucleic acid types differ in the structure of the sugar in their nucleotides–DNA contains 2'-deoxyribose while RNA contains ribose (where the only difference is the presence of a hydroxyl group). Also, the nucleobases found in the two nucleic acid types are different: adenine, cytosine, and guanine are found in both RNA and DNA, while thymine occurs in DNA and uracil occurs in RNA.

The sugars and phosphates in nucleic acids are connected to each other in an alternating chain (sugar-phosphate backbone) through phosphodiester linkages. In conventional nomenclature, the carbons to which the phosphate groups attach are the 3'-end and the 5'-end carbons of the sugar. This gives nucleic acids directionality, and the ends of nucleic acid molecules are referred to as 5'-end and 3'-end. The nucleobases are joined to the sugars via an N-glycosidic linkage involving a nucleobase ring nitrogen (N-1 for pyrimidines and N-9 for purines) and the 1' carbon of the pentose sugar ring.

Non-standard nucleosides are also found in both RNA and DNA and usually arise from modification of the standard nucleosides within the DNA molecule or the primary (initial) RNA transcript. Transfer RNA (tRNA) molecules contain a particularly large number of modified nucleosides.

Topology

Double-stranded nucleic acids are made up of complementary sequences, in which extensive Watson-Crick base pairing results in a highly repeated and quite uniform double-helical three-dimensional structure. In contrast, single-stranded RNA and DNA molecules are not constrained to a regular double helix, and can adopt highly complex three-dimensional structures that are based on short stretches of intramolecular base-paired sequences including both Watson-Crick and noncanonical base pairs, and a wide range of complex tertiary interactions.

Nucleic acid molecules are usually unbranched, and may occur as linear and circular molecules. For example, bacterial chromosomes, plasmids, mitochondrial DNA, and chloroplast DNA are usually circular double-stranded DNA molecules, while chromosomes of the eukaryotic nucleus are usually linear double-stranded DNA molecules. Most RNA molecules are linear, single-stranded molecules, but both circular and branched molecules can result from RNA splicing reactions.

Nucleic Acid Sequences

One DNA or RNA molecule differs from another primarily in the sequence of nucleotides. Nucleotide sequences are of great importance in biology since they carry the ultimate instructions that encode all biological molecules, molecular assemblies, subcellular and cellular structures, organs, and organisms, and directly enable cognition, memory, and behavior (See: Genetics). Enormous efforts have gone into the development of experimental methods to determine the nucleotide sequence of biological DNA and RNA molecules, and today hundreds of millions of nucleotides are sequenced daily at genome centers and smaller laboratories worldwide. In addition to maintaining the GenBank nucleic acid sequence database, the National Center for Biotechnology Information (NCBI, http://www.ncbi.nlm.nih.gov) provides analysis and retrieval resources for the data in GenBank and other biological data made available through the NCBI Web site.

Types of Nucleic Acids

Deoxyribonucleic Acid

Deoxyribonucleic acid (DNA) is a nucleic acid containing the genetic instructions used in the development and functioning of all known living organisms. The DNA segments carrying this genetic information are called genes. Likewise, other DNA sequences have structural purposes, or are involved in regulating the use of this genetic information. Along with RNA and proteins, DNA is one of the three major macromolecules that are essential for all known forms of life. DNA consists of two long polymers of simple units called nucleotides, with backbones made of sugars and phosphate groups joined by ester bonds. These two strands run in opposite directions to each other and are, therefore, anti-parallel. Attached to each sugar is one of four types of molecules called nucleobases (informally, bases). It is the sequence of these four nucleobases along the backbone that encodes information. This information is read using the genetic code, which specifies the sequence of the amino acids within proteins. The code is read by copying stretches of DNA into the related nucleic acid RNA in a process called transcription. Within cells DNA is organized into long structures called chromosomes. During cell division these chromosomes are duplicated in the process of DNA replication, providing each cell its own complete set of chromosomes. Eukaryotic organisms (animals, plants, fungi, and protists) store most of their DNA inside the cell nucleus and some of their DNA in organelles, such as mitochondria or chloroplasts. In contrast, prokaryotes (bacteria and archaea) store their DNA only in the cytoplasm. Within the chromosomes, chromatin proteins such as histones compact and organize DNA. These compact structures guide the interactions between DNA and other proteins, helping control which parts of the DNA are transcribed.

Ribonucleic Acid

Ribonucleic acid (RNA) functions in converting genetic information from genes into the amino acid sequences of proteins. The three universal types of RNA include transfer RNA (tRNA), messenger RNA (mRNA), and ribosomal RNA (rRNA). Messenger RNA acts to carry genetic sequence information between DNA and ribosomes, directing protein synthesis. Ribosomal RNA is a major component of the ribosome, and catalyzes peptide bond formation. Transfer RNA serves as the carrier molecule for amino acids to be used in protein synthesis, and is responsible for decoding the mRNA. In addition, many other classes of RNA are now known.

Artificial Nucleic Acid

Artificial nucleic acid analogues have been designed and synthesized by chemists, and include peptide nucleic acid, morpholino- and locked nucleic acid, glycol nucleic acid, and threose nucleic acid. Each of these is distinguished from naturally occurring DNA or RNA by changes to the backbone of the molecules.

NUCLEIC ACID ANALOGUE

Nucleic acid analogues are compounds which are analogous (structurally similar) to naturally occurring RNA and DNA, used in medicine and in molecular biology research. Nucleic acids are chains of nucleotides, which are composed of three parts: a phosphate backbone, a pentose sugar, either ribose or deoxyribose, and one of four nucleobases. An analogue may have any of these altered. Typically the analogue nucleobases confer, among other things, different base pairing and base stacking properties. Examples include universal bases, which can pair with all four canonical bases, and phosphate-sugar backbone analogues such as PNA, which affect the properties of the chain (PNA can even form a triple helix). Nucleic acid analogues are also called Xeno Nucleic Acid and represent one of the main pillars of xenobiology, the design of new-to-nature forms of life based on alternative biochemistries.

Artificial nucleic acids include peptide nucleic acid (PNA), Morpholino and locked nucleic acid (LNA), as well as glycol nucleic acid (GNA) and threose nucleic acid (TNA). Each of these is distinguished from naturally occurring DNA or RNA by changes to the backbone of the molecule.

In May 2014, researchers announced that they had successfully introduced two new artificial nucleotides into bacterial DNA, and by including individual artificial nucleotides in the culture media, were able to passage the bacteria 24 times; they did not create mRNA or proteins able to use the artificial nucleotides. The artificial nucleotides featured 2 fused aromatic rings.

Medicine

Several nucleoside analogues are used as antiviral or anticancer agents. The viral polymerase incorporates these compounds with non-canonical bases. These compounds are activated in the cells by being converted into nucleotides, they are administered as nucleosides since charged nucleotides cannot easily cross cell membranes.

Molecular Biology

Nucleic acid analogues are used in molecular biology for several purposes: Investigation of possible scenarios of the origin of life: By testing different analogs, researchers try to answer the question of whether life's use of DNA and RNA was selected over time due to its advantages, or if they were chosen by arbitrary chance; As a tool to detect particular sequences: XNA can be used to tag and identify a wide range of DNA and RNA components with high specificity and accuracy; As an enzyme acting on DNA, RNA and XNA substrates - XNA has been shown to have the ability to cleave and ligate DNA, RNA and other XNA molecules similar to the actions of RNA ribozymes; As a tool with resistance to RNA hydrolysis; Investigation of the mechanisms used by enzyme; Investigation of the structural features of nucleic acids.

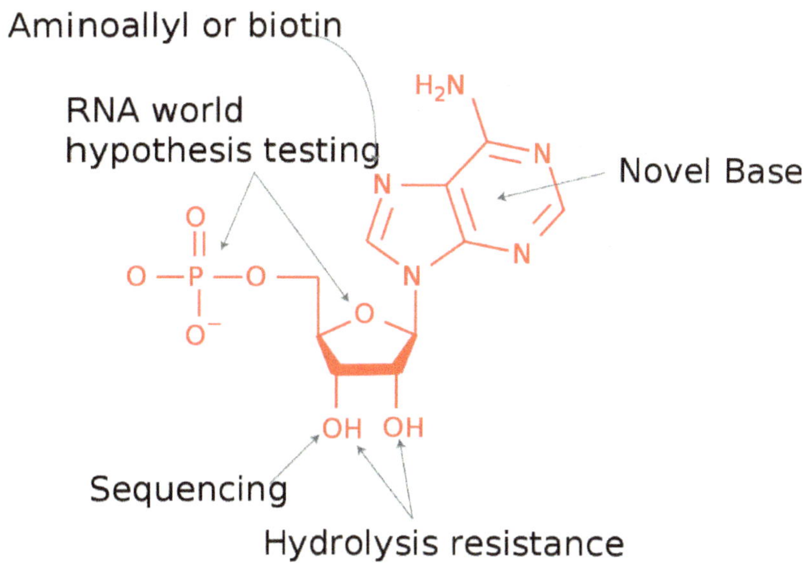

Common changes in nucleotide analogues

BACKBONE ANALOGUES

Hydrolysis Resistant RNA-Analogues

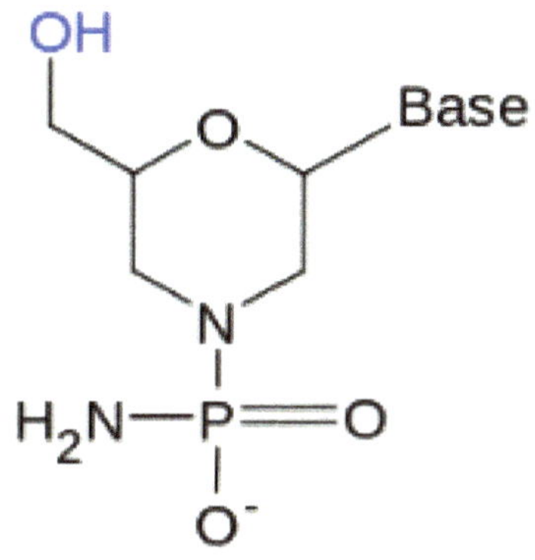

Chemical structure of Morpholino

To overcome the fact that ribose's 2' hydroxy group that reacts with the phosphate linked 3' hydroxy group (RNA is too unstable to be used or synthesized reliably), a ribose analogue is used. The most common RNA analogues are 2'-O-methyl-substituted RNA, locked nucleic acid (LNA) or bridged nucleic acid (BNA), morpholino, and peptide nucleic acid (PNA). Although these oligonucleotides have a different backbone sugar or, in the case of PNA, an amino acid residue in place of the ribose phosphate, they still bind to RNA or DNA according to Watson and Crick pairing, but are immune to nuclease activity. They cannot be synthesized enzymatically and can only be obtained synthetically using phosphoramidite strategy or, for PNA, methods of peptide synthesis.

Other Notable Analogues Used as Tools

Dideoxynucleotides are used in sequencing . These nucleoside triphosphates possess a non-canonical sugar, dideoxyribose, which lacks the 3› hydroxyl group normally present in DNA and therefore cannot bond with the next base. The lack of the 3› hydroxyl group terminates the chain reaction as the DNA polymerases mistake it for a regular deoxyribonucleotide. Another chain-terminating analogue that lacks a 3› hydroxyl and mimics adenosine is called cordycepin. Cordycepin is an anticancer drug that targets RNA replication. Another analogue in sequencing is a nucleobase analogue, 7-deaza-GTP and is used to sequence CG rich regions, instead 7-deaza-ATP is called tubercidin, an antibiotic.

Precursors to the RNA World

RNA may be too complex to be the first nucleic acid, so before the RNA world several simpler nucleic acids that differ in the backbone, such as TNA and GNA and PNA, have been offered as candidates for the first nucleic acids.

Base Analogues

Nucleobase Structure and Nomenclature

For structures of the analogues that may be mentioned in the literature, see Simple aromatic ring.

Naturally occurring bases can be divided into two classes according to their structure:

- pyrimidines are six-membered heterocyclic with nitrogen atoms in position 1 and 3.
- purines are bicyclic, consisting of a pyrimidine fused to an imidazole ring.

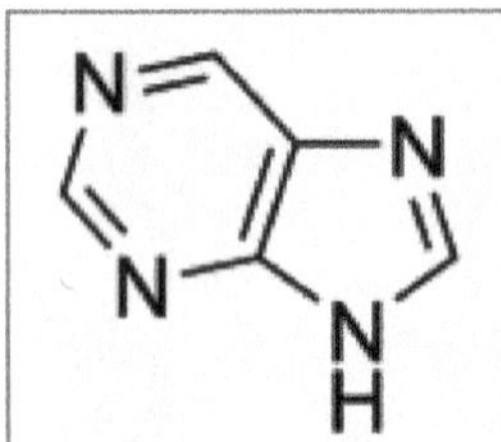	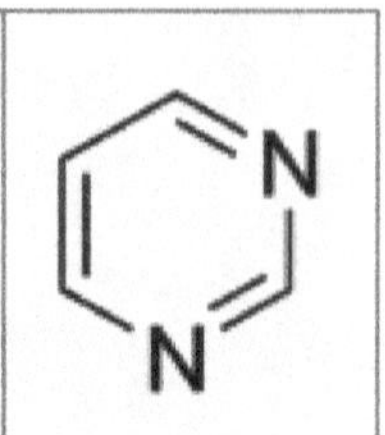
Purine	Pyrimidine

Artificial nucleotides (*Unnatural Base Pairs* (UBPs) named *d5SICS UBP* and *dNaM UBP*) have been inserted into bacterial DNA but these genes did not template mRNA or induce protein synthesis. The artificial nucleotides featured two fused aromatic rings which formed a (d5SICS–dNaM) complex mimicking the natural (dG–dC) base pair.

Fluorophores

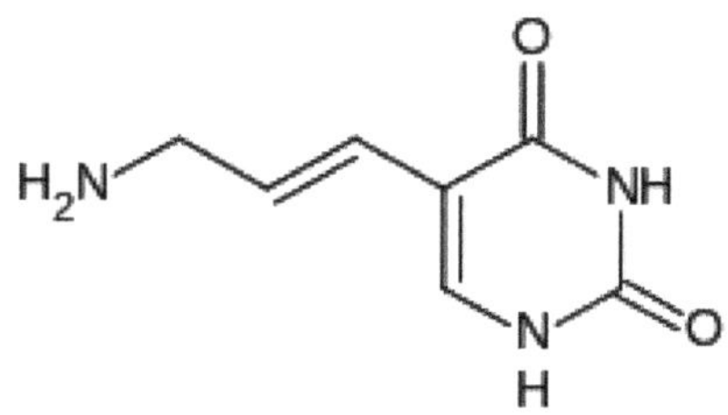

Structure of aminoallyl-uridine

Commonly fluorophores (such as rhodamine or fluorescein) are linked to the ring linked to the sugar (in para) via a flexible arm, presumably extruding from the major groove of the helix. Due to low processivity of the nucleotides linked to bulky adducts such as florophores by taq polymerases, the sequence is typically copied using a nucleotide with an arm and later coupled with a reactive fluorophore (indirect labelling):

- amine reactive: Aminoallyl nucleotide contain a primary amine group on a linker that reacts with the amino-reactive dye such as a cyanine or Alexa Fluor dyes, which contain a reactive leaving group, such as a succinimidyl ester (NHS). (base-pairing amino groups are not affected).
- thiol reactive: thiol containing nucleotides reacts with the fluorophore linked to a reactive leaving group, such as a maleimide.
- biotin linked nucleotides rely on the same indirect labelling principle (+ fluorescent streptavidin) and are used in Affymetrix DNAchips.

Fluorophores find a variety of uses in medicine and biochemistry.

Fluorescent Base Analogues

The most commonly used and commercially available fluorescent base analogue, 2-aminopurine (2-AP), has a high-fluorescence quantum yield free in solution (0.68) that is considerably reduced (appr. 100 times but highly dependent on base sequence) when incorporated into nucleic acids. The emission sensitivity of 2-AP to immediate surroundings is shared by other promising and useful fluorescent base analogues like 3-MI, 6-MI, 6-MAP, pyrrolo-dC (also commercially available),modified and improved derivatives of pyrrolo-dC, furan-modified bases and many other ones (see recent reviews). This sensitivity to the microenvironment has been utilized in studies of e.g. structure and dynamics within both DNA and RNA, dynamics and kinetics of DNA-protein interaction and electron transfer within DNA. A newly developed and very interesting group of fluorescent base analogues that has a fluorescence quantum yield that is nearly insensitive to their immediate surroundings is the tricyclic cytosine family. 1,3-Diaza-

2-oxophenothiazine, tC, has a fluorescence quantum yield of approximately 0.2 both in single- and in double-strands irrespective of surrounding bases. Also the oxo-homologue of tC called tC^O (both commercially available), 1,3-diaza-2-oxophenoxazine, has a quantum yield of 0.2 in double-stranded systems. However, it is somewhat sensitive to surrounding bases in single-strands (quantum yields of 0.14–0.41). The high and stable quantum yields of these base analogues make them very bright, and, in combination with their good base analogue properties (leaves DNA structure and stability next to unperturbed), they are especially useful in fluorescence anisotropy and FRET measurements, areas where other fluorescent base analogues are less accurate. Also, in the same family of cytosine analogues, a FRET-acceptor base analogue, tC_{nitro}, has been developed. Together with tC^O as a FRET-donor this constitutes the first nucleic acid base analogue FRET-pair ever developed. The tC-family has, for example, been used in studies related to polymerase DNA-binding and DNA-polymerization mechanisms.

Natural Non-Canonical Bases

In a cell, there are several non-canonical bases present: CpG islands in DNA (are often methylated), all eukaryotic mRNA (capped with a methyl-7-guanosine), and several bases of rRNAs (are methylated). Often, tRNAs are heavily modified postranscriptionally in order to improve their conformation or base pairing, in particular in/near the anticodon: inosine can base pair with C, U, and even with A, whereas thiouridine (with A) is more specific than uracil (with a purine). Other common tRNA base modifications are pseudouridine (which gives its name to the TΨC loop), dihydrouridine (which does not stack as it is not aromatic), queuosine, wyosine, and so forth. Nevertheless, these are all modifications to normal bases and are not placed by a polymerase.

Base-Pairing

Canonical bases may have either a carbonyl or an amine group on the carbons surrounding the nitrogen atom furthest away from the glycosidic bond, which allows them to base pair (Watson-Crick base pairing) via hydrogen bonds (amine with ketone, purine with pyrimidine). Adenine and 2-aminoadenine have one/two amine group(s), whereas thymine has two carbonyl groups, and cytosine and guanine are mixed amine and carbonyl (inverted in respect to each other).

Natural basepairs

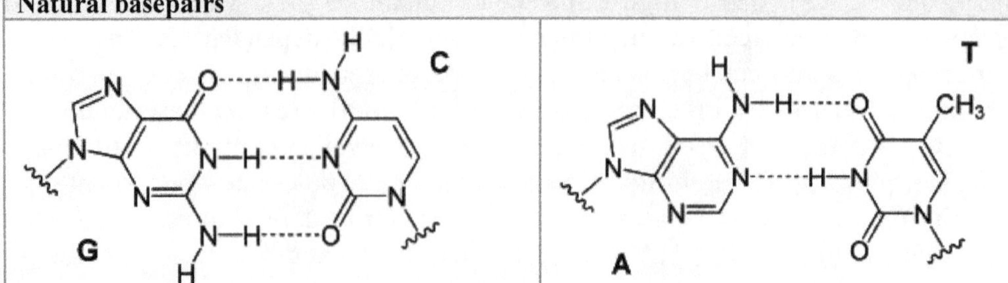

A GC basepair: purine carbonyl/amine forms three intermolecular hydrogen bonds with pyrimidine amine/carbonyl	An AT basepair: purine amine/- forms two intermolecular hydrogen bonds with pyrimidine carbonyl/carbonyl

The precise reason why there are only four nucleotides is debated, but there are several unused possibilities. Furthermore, adenine is not the most stable choice for base pairing: in Cyanophage S-2L diaminopurine (DAP) is used instead of adenine (host evasion). Diaminopurine basepairs perfectly with thymine as it is identical to adenine but has an amine group at position 2 forming 3 intramolecular hydrogen bonds, eliminating the major difference between the two types of basepairs (Weak:A-T and Strong:C-G). This improved stability affects protein-binding interactions that rely on those differences. Other combination include,

- isoguanine and isocytosine, which have their amine and ketone inverted compared to standard guanine and cytosine, (not used probably as tautomers are problematic for base pairing, but isoC and isoG can be amplified correctly with PCR even in the presence of the 4 canonical bases)
- diaminopyrimidine and a xanthine, which bind like 2-aminoadenine and thymine but with inverted structures (not used as xanthine is a deamination product)

Unused basepair arrangements		
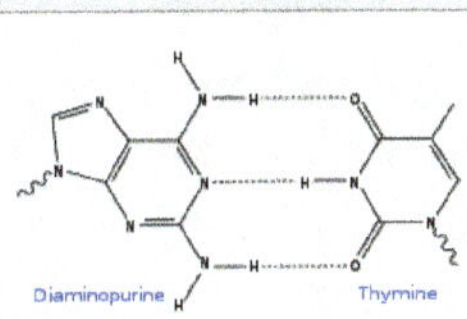	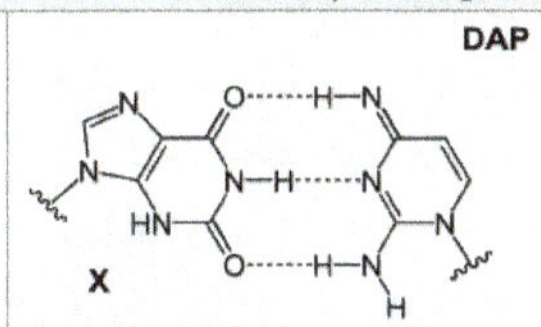	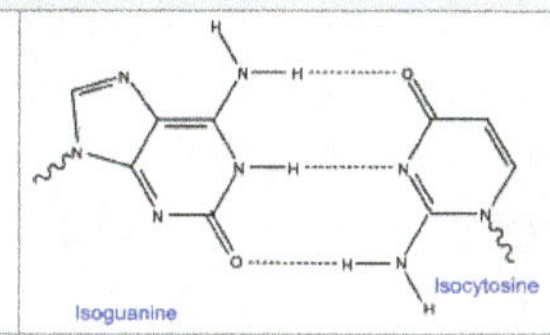
A DAP-T base: purine amine/amine forms three intermolecular hydrogen bonds with pyrimidine ketone/ketone	An X-DAP base: purine ketone/ketone forms three intermolecular hydrogen bonds with pyrimidine amine/amine	A iG-iC base: purine amine/ketone forms three intermolecular hydrogen bonds with pyrimidine ketone/amine

However, correct DNA structure can form even when the bases are not paired via hydrogen bonding; that is, the bases pair thanks to hydrophobicity, as studies have shown using DNA isosteres (analogues with same number of atoms), such as the thymine analogue 2,4-difluorotoluene (F) or the adenine analogue 4-methylbenzimidazole (Z). An alternative hydrophobic pair could be isoquinoline, and the pyrrolo[2,3-b]pyridine.

Other noteworthy basepairs:

- Several fluorescent bases have also been made, such as the 2-amino-6-(2-thienyl) purine and pyrrole-2-carbaldehyde base pair.
- Metal coordinated bases, such as two 2,6-bis(ethylthiomethyl)pyridine (SPy) with a silver ion or pyridine-2,6-dicarboxamide (Dipam) and a mondentate pyridine (Py) with a copper ion.
- Universal bases may pair indiscriminately with any other base, but, in general, lower the melting temperature of the sequence considerably; examples include 2›-deoxyinosine (hypoxanthine deoxynucleotide) derivatives, nitroazole analogues, and hydrophobic aromatic non-hydrogen-bonding bases (strong stacking effects). These are used as proof of concept and, in general, are not utilised in degenerate primers (which are a mixture of primers).

- The numbers of possible base pairs is doubled when xDNA is considered. xDNA contains expanded bases, in which a benzene ring has been added, which may pair with canonical bases, resulting in four possible base-pairs (8 bases:xA-T,xT-A,xC-G,xG-C, 16 bases if the unused arrangements are used). Another form of benzene added bases is yDNA, in which the base is widened by the benzene.

Novel basepairs with special properties		
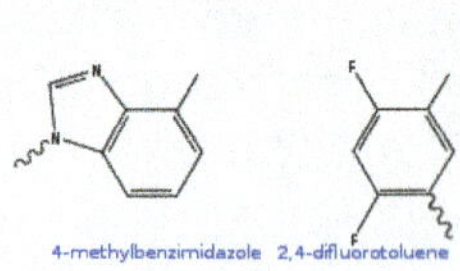	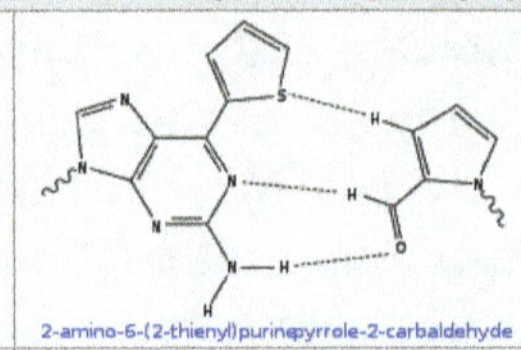	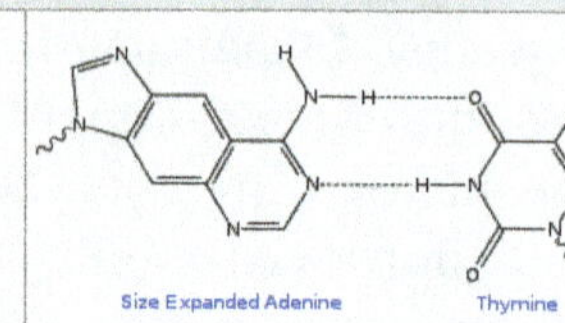
A F-Z base: methylbenzimidazole does not form intermolecular hydrogen bonds with toluene F/F	An S-Pa base: purine thienyl/amine forms three intermolecular hydrogen bonds with pyrrole -/carbaldehyde	An xA-T base: same bonding as A-T

Metal Base-Pairs

In metal base-pairing, the Watson-Crick hydrogen bonds are replaced by the interaction between a metal ion with nucleosides acting as ligands. The possible geometries of the metal that would allow for duplex formation with two bidentate nucleosides around a central metal atom are: tetrahedral, dodecahedral, and square planar. Metal-complexing with DNA can occur by the formation of non-canonical base pairs from natural nucleobases with participation by metal ions and also by the exchanging the hydrogen atoms that are part of the Watson-Crick base pairing by metal ions. Introduction of metal ions into a DNA duplex has shown to have potential magnetic, conducting properties, as well as increased stability.

Metal complexing has been shown to occur between natural nucleobases. A well-documented example is the formation of T-Hg-T, which involves two deprotonated thymine nucleobases that are brought together by Hg^{2+} and forms a connected metal-base pair. This motif does not accommodate stacked Hg^{2+} in a duplex due to an intrastrand hairpin formation process that is favored over duplex formation. Two thymines across from each other in a duplex do not form a Watson-Crick base pair in a duplex; this is an example where a Watson-Crick basepair mismatch is stabilized by the formation of the metal-base pair. Another example of a metal complexing to natural nucleobases is the formation of A-Zn-T and G-Zn-C at high pH; Co^{+2} and Ni^{+2} also form these complexes. These are Watson-Crick base pairs where the divalent cation in coordinated to the nucleobases. The exact binding is debated.

A large variety of artificial nucleobases have been developed for use as metal base pairs. These modified nucleobases exhibit tunable electronic properties, sizes, and binding affinities that can be optimized for a specific metal. For, example a nucleoside modified with a pyridine-2,6-dicarboxylate has shown to bind tightly to Cu^{2+}, whereas other divalent ions are only loosely bound. The tridentate character contributes to this selectivity. The fourth coordination site on the copper is saturated by an oppositely arranged pyridine nucleobase. The asymmetric metal base pairing system is orthogonal to the Watson-Crick base pairs. Another example of an artificial nucleobase is that with hydroxypyridone nucleobases, which are able to bind Cu^{2+} inside the DNA duplex. Five consecutive copper-hydroxypyridone base pairs were incorporated into a double strand, which were flanked by only one natural nucleobase on both ends. EPR data showed that the distance between copper centers was estimated to be 3.7 ± 0.1 Å, while a natural B-type DNA duplex is only slightly larger (3.4 Å). The appeal for stacking metal ions inside a DNA duplex is the hope to obtain nanoscopic self-assembling metal wires, though this has not been realized yet.

Unnatural Base Pair (UBP)

An unnatural base pair (UBP) is a designed subunit (or nucleobase) of DNA which is created in a laboratory and does not occur in nature. In 2012, a group of American scientists led by Floyd Romesberg, a chemical biologist at the Scripps Research Institute in San Diego, California, published that his team designed an unnatural base pair (UBP). The two new artificial nucleotides or *Unnatural Base Pair* (UBP) were named d5SICS and dNaM. More technically, these artificial nucleotides bearing hydrophobic nucleobases, feature two fused aromatic rings that form a (d5SICS–dNaM) complex or base pair in DNA. In 2014 the same team from the Scripps Research Institute reported that they synthesized a stretch of circular DNA known as a plasmid containing natural T-A and C-G base pairs along with the best-performing UBP Romesberg›s laboratory had designed, and inserted it into cells of the common bacterium *E. coli* that successfully replicated the unnatural base pairs through multiple generations. This is the first known example of a living organism passing along an expanded genetic code to subsequent generations. This was in part achieved by the addition of a supportive algal gene that expresses a nucleotide triphosphate transporter which efficiently imports the triphosphates of both d5SICSTP and dNaMTP into *E. coli* bacteria. Then, the natural bacterial replication pathways use them to accurately replicate the plasmid containing d5SICS–dNaM.

The successful incorporation of a third base pair is a significant breakthrough toward the goal of greatly expanding the number of amino acids which ean be encoded by DNA, from the existing 20 amino acids to a theoretically possible 172, thereby expanding the potential for living organisms to produce novel proteins. The artificial strings of DNA do not encode for anything yet, but scientists speculate they could be designed to manufacture new proteins which could have industrial or pharmaceutical uses.

Another demonstration of UBPs were achieved by Ichiro Hirao›s group at RIKEN institute in Japan. In 2002, they developed an unnatural base pair between 2-amino-8-(2-thienyl)purine (s) and pyridine-2-one (y) that functions *in vitro* in transcription and translation, for the site-specific incorporation of non-standard amino acids into proteins. In 2006, they created 7-(2-thienyl)imidazo[4,5-b]pyridine (Ds) and pyrrole-2-carbaldehyde (Pa) as a third base pair for replication and transcription.Afterward, Ds and 4-[3-(6-aminohexanamido)-1-propynyl]-2-nitropyrrole (Px) was discovered as a high fidelity pair in PCR amplification. In 2013, they applied the Ds-Px pair to DNA aptamer generation by *in vitro* selection (SELEX) and demonstrated the genetic alphabet expansion significantly augment DNA aptamer affinities to target proteins.

Orthogonal System

The possibility has been proposed and studied, both theoretically and experimentally, of implementing an orthogonal system inside cells independent of the cellular genetic material in order to make a completely safe system, with the possible increase in encoding potentials Several groups have focused on different aspects:

- novel backbones and base pairs as discussed above
- XNA (Xeno Nucleic Acid) artificial replication/transcription polymerases starting generally from T7 RNA polymerase
- ribosomes (16S sequences with altered anti Shine-Dalgarno sequence allowing the translation of only orthogonal mRNA with a matching altered Shine-Dalgarno sequence)
- novel tRNA encoding non-natural aminoacids. See Expanded genetic code

REFERENCES

1. He called them nuclein.
2. Bill Bryson, A Short History of Nearly Everything, Broadway Books, 2005, p. 500.
3. "What is DNA". What is DNA. Linda Clarks. Retrieved 6 August 2016.
4. Dahm, R (January 2008). "Discovering DNA: Friedrich Miescher and the early years of nucleic acid research". Human Genetics. 122 (6): 565–81. doi:10.1007/s00439-007-0433-0. ISSN 0340-6717. PMID 17901982.
5. Cox, Michael; Nelson, David (2008). Principles of Biochemistry. Susan Winslow. p. 288.
6. "DNA Structure". What is DNA. Linda Clarks. Retrieved 6 August 2016.
7. International Human Genome Sequencing Consortium (2001). "Initial sequencing and analysis of the human genome." (PDF). Nature. 409 (6822): 860–921. doi:10.1038/35057062. PMID 11237011.
8. Venter, JC; et al. (2001). "The sequence of the human genome." (PDF). Science. 291 (5507): 1304–1351. Bibcode:2001Sci...291.1304V. doi:10.1126/science.1058040. PMID 11181995.
9. Budowle B, van Daal A (April 2009). "Extracting evidence from forensic DNA analyses: future molecular biology directions". BioTechniques. 46 (5): 339–40, 342–50. doi:10.2144/000113136. PMID 19480629.
10. Alberts, Bruce (2008). Molecular biology of the cell. New York: Garland Science. ISBN 0-8153-4105-9.
11. Elson D (1965). "Metabolism of nucleic acids (macromolecular DNA and RNA)". Annu. Rev. Biochem. 34: 449–86. doi:10.1146/annurev.bi.34.070165.002313. PMID 14321176.
12. National Institute of Health (September 28, 2007). "Discovering DNA: Friedrich Miescher and the early years of nucleic acid research.". Hum. Genet. nih.gov.

122: 565–81. doi:10.1007/s00439-007-0433-0. PMID 17901982.

13. Aparadh, V. T. & B. A. Karadge (2012). "Infrared Spectroscopic Studies in Some Cleome species" (PDF). ISSN 2319-8877.
14. Brock, Thomas D.; Madigan, Michael T. (2009). Brock biology of microorganisms. Pearson / Benjamin Cummings. ISBN 0-321-53615-0.
15. Hardinger, Steven; University of California, Los Angeles (2011). "Knowing Nucleic Acids" (PDF). ucla.edu.
16. Mullis, Kary B. The Polymerase Chain Reaction (Nobel Lecture). 1993. (retrieved December 1, 2010) http://nobelprize.org/nobel_prizes/chemistry/laureates/1993/mullis-lecture.html
17. Verma S, Eckstein F (1998). "Modified oligonucleotides: synthesis and strategy for users". Annu. Rev. Biochem. 67: 99–134. doi:10.1146/annurev.biochem.67.1.99. PMID 9759484.
18. Stryer, Lubert; Berg, Jeremy Mark; Tymoczko, John L. (2007). Biochemistry. San Francisco: W.H. Freeman. ISBN 0-7167-6766-X.
19. Gregory SG, Barlow KF, McLay KE, et al. (May 2006). "The DNA sequence and biological annotation of human chromosome 1". Nature. 441 (7091): 315–21. Bibcode:2006Natur.441..315G. doi:10.1038/nature04727. PMID 16710414.
20. Todorov TI, Morris MD (April 23, 2002). National Institutes of Health. "Comparison of RNA, single-stranded DNA and double-stranded DNA behavior during capillary electrophoresis in semidilute polymer solutions". Electrophoresis. nih.gov. 23: 1033–44. doi:10.1002/1522-2683(200204)23:7/8<1033::AID-ELPS1033>3.0.CO;2-7. PMID 11981850.
21. Margaret Hunt; University of South Carolina (2010). "RN Virus Replication Strategies". sc.edu.
22. McGlynn Peter; Robert G. Lloyd (June 10, 1999). "RecG helicase activity at three- and four-strand DNA structures". oxfordjournals.org. ISSN 1362-4962.
23. Rich A, RajBhandary UL (1976). "Transfer RNA: molecular structure, sequence, and properties". Annu. Rev. Biochem. 45: 805–60. doi:10.1146/annurev.bi.45.070176.004105. PMID 60910.
24. Watson JD, Crick FH (April 1953). "Molecular structure of nucleic acids; a structure for deoxyribose nucleic acid". Nature. 171 (4356): 737–8. Bibcode:1953Natur.171..737W. doi:10.1038/171737a0. PMID 13054692.
25. Ferré-D'Amaré AR, Doudna JA (1999). "RNA folds: insights from recent crystal structures". Annu Rev Biophys Biomol Struct. 28: 57–73. doi:10.1146/annurev.biophys.28.1.57. PMID 10410795.
26. Gilbert, Walter G. 1980. DNA Sequencing and Gene Structure (Nobel Lecture) http://nobelprize.org/nobel_prizes/chemistry/laureates/1980/gilbert-lecture.html
27. Sanger, Frederick. 1980. Determination of Nucleotide Sequences in DNA (Nobel Lecture) http://nobelprize.org/nobel_prizes/chemistry/laureates/1980/sanger-lecture.html
28. "Database resources of the National Center for Biotechnology Information".

Nucleic Acids Research. 42: D7–17. 2014. doi:10.1093/nar/gkt1146. PMC 3965057 . PMID 24259429.

29. Singer, Emily (July 19, 2015). "Chemists Invent New Letters for Nature's Genetic Alphabet". Wired (magazine). Retrieved July 20, 2015.
30. Petersson, B; et al. (Feb 2005). "Crystal structure of a partly self-complementary peptide nucleic acid (PNA) oligomer showing a duplex-triplex network". J Am Chem Soc. 127 (5): 1424–30. doi:10.1021/ja0458726.
31. Taylor, Alexander I.; Pinheiro, Vitor B.; Smola, Matthew J.; Morgunov, Alexey S.; Peak-Chew, Sew; Cozens, Christopher; Weeks, Kevin M.; Herdewijn, Piet; Holliger, Philipp. "Catalysts from synthetic genetic polymers". Nature. 518 (7539): 427–430. doi:10.1038/nature13982. PMC 4336857 . PMID 25470036.
32. Wang, Qi; Chen, Lei; Long, Yitao; Tian, He; Wu, Junchen. "Molecular Beacons of Xeno-Nucleic Acid for Detecting Nucleic Acid". Theranostics. 3 (6): 395–408. doi:10.7150/thno.5935. PMC 3677410 . PMID 23781286.
33. Summerton, J; Weller, D (1997). "Morpholino Antisense Oligomers: Design, Preparation and Properties". Antisense & Nucleic Acid Drug Development. 7: 187–195. doi:10.1089/oli.1.1997.7.187. PMID 9212909.
34. Summerton, J (1999). "Morpholino Antisense Oligomers: The Case for an RNase-H Independent Structural Type". Biochimica et Biophysica Acta. 1489: 141–158. doi:10.1016/s0167-4781(99)00150-5.
35. Pollack, Andrew (May 7, 2014). "Researchers Report Breakthrough in Creating Artificial Genetic Code". New York Times. Retrieved May 7, 2014.
36. Callaway, Ewen (May 7, 2014). "First life with 'alien' DNA". Nature. doi:10.1038/nature.2014.15179. Retrieved May 7, 2014.
37. Malyshev, Denis A.; Dhami, Kirandeep; Lavergne, Thomas; Chen, Tingjian; Dai, Nan; Foster, Jeremy M.; Corrêa, Ivan R.; Romesberg, Floyd E. (May 7, 2014). "A semi-synthetic organism with an expanded genetic alphabet". Nature. 509: 385–388. doi:10.1038/nature13314. PMC 4058825 . PMID 24805238. Retrieved May 7, 2014.
38. Ward; et al. (1969). "Fluorescence Studies of Nucleotides and Polynucleotides I. Formycin 2-Aminopurine Riboside 2,6-Diaminopurine Riboside and Their Derivatives". J. Biol. Chem. 244: 1228–37.
39. Hawkins (2001). "Fluorescent pteridine nucleoside analogs - A window on DNA interactions". Cell Biochem. Biophys. 34: 257–81. doi:10.1385/cbb:34:2:257.
40. Berry; et al. (2004). "Pyrrolo-dC and pyrrolo-C: fluorescent analogs of cytidine and 2 '-deoxycytidine for the study of oligonucleotides". Tetrahedron Lett. 45: 2457–61. doi:10.1016/j.tetlet.2004.01.108.
41. Wojciechowski; et al. (2008). "Fluorescence and hybridization properties of peptide nucleic acid containing a substituted phenylpyrrolo-cytosine designed to engage guanine with an additional H-bond". J. Am. Chem. Soc. 130: 12574–12575. doi:10.1021/ja804233g.
42. Greco; et al. (2005). "Simple fluorescent pyrimidine analogues detect the presence

of DNA abasic sites". J. Am. Chem. Soc. 127: 10784–85. doi:10.1021/ja052000a.

43. Rist; et al. (2002). "Fluorescent nucleotide base analogs as probes of nucleic acid structure, dynamics and interactions". Curr. Org. Chem. 6: 775–93. doi:10.2174/1385272023373914.
44. Wilson; et al. (2006). "Fluorescent DNA base replacements: reporters and sensors for biological systems". Org, & Biomol. Chem. 4: 4265–74. doi:10.1039/b612284c.
45. Wilhelmsson and Tor (2016). Fluorescent Analogs of Biomolecular Building Blocks: Design and Applications. New Jersey: Wiley. ISBN 978-1-118-17586-6.
46. Wilhelmsson (2010). "Fluorescent nucleic acid base analogues". Q. Rev. Biophys. 43 (2): 159–183. doi:10.1017/s0033583510000090.
47. Sinkeldam; et al. (2010). "Fluorescent analogs of Biomolecular Building Blocks: Design, properties and applications". Chem. Rev. 110 (5): 2579–2619. doi:10.1021/cr900301e.
48. Wilhelmsson; et al. (2001). "A highly fluorescent DNA base analogue that forms Watson-Crick base pairs with guanine". J. Am. Chem. Soc. 123: 2434–35. doi:10.1021/ja0025797.
49. Sandin; et al. (2005). "Fluorescent properties of DNA base analogue tC upon incorporation into DNA - negligible influence of neighbouring bases on fluorescence quantum yield". Nucleic Acids Res. 33: 5019–25. doi:10.1093/nar/gki790.
50. Sandin; et al. (2008). "Characterization and use of an unprecedentedly bright and structurally non-perturbing fluorescent DNA base analogue". Nucleic Acids Res. 36: 157–67. doi:10.1093/nar/gkm1006.
51. Börjesson; et al. (2009). "Nucleic acid base analog FRET-pair facilitating detailed structural measurements in nucleic acid containing systems". J. Am. Chem. Soc. 131: 4288–93. doi:10.1021/ja806944w.
52. ^ Jump up to:a b Rodriguez, Hernandez (2013). "Structural and mechanistic basis for enhanced translational efficiency by 2-thiouridine at the tRNA anticodon wobble position.". J. Mol. Biol. 425: 3888–906. doi:10.1016/j.jmb.2013.05.018. PMC 4521407 . PMID 23727144.
53. Kirnos MD, Khudyakov IY, Alexandrushkina NI, Vanyushin BF. 2-aminoadenine is an adenine substituting for a base in S-2L cyanophage DNA. Nature. 1977 Nov 24;270(5635):369–70.
54. Johnson SC et al. A third base pair for the polymerase chain reaction: inserting isoC and isoG. Nucleic Acids Res. 2004 Mar 29;32(6):1937–41.
55. Taniguchi, Y; Kool, ET (Jul 2007). "Nonpolar isosteres of damaged DNA bases: effective mimicry of mutagenic properties of 8-oxopurines". J Am Chem Soc. 129 (28): 8836–44. doi:10.1021/ja071970q.
56. Hwang, G. T.; Romesberg, F. E. (2008). "Unnatural Substrate Repertoire of A, B, and X Family DNA Polymerases". J. Am. Chem. Soc. 130: 14872–14882. doi:10.1021/ja803833h.

57. Kimoto, M; et al. (2007). "Fluorescent probing for RNA molecules by an unnatural base-pair system". Nucleic Acids Res. 35 (16): 5360–9. doi:10.1093/nar/gkm508.

58. Zimmermann, N; et al. (Feb 2004). "A second-generation copper(II)-mediated metallo-DNA-base pair". Bioorg Chem. 32 (1): 13–25. doi:10.1016/j.bioorg.2003.09.001.

59. Liu, H; et al. (2003). "A four-base paired genetic helix with expanded size". Science. 302 (5646): 868–71. doi:10.1126/science.1088334.

60. Wettig, S. D.; Lee, J.S. (2003). J. Inorg. Biochem. 94: 94–99.Missing or empty |title= (help)

61. Zhang, H.; Calzolari, A.; Felice, R. Di (2005). "On the Magnetic Alignment of Metal Ions in a DNA-Mimic Double Helix". J. Phys. Chem. B. 109: 15345–15348. doi:10.1021/jp052202t.

62. Aich, P.; Skinner, R. J. S.; Wettig, S. D.; Steer, R. P.; Lee, J. S. (2002). Biomol. Struct. Dyn. 20: 93–98. Missing or empty |title= (help)

63. Clever, G. H.; Polborn, K.; Carell, T. (2005). "Ein hochgradig DNA-Duplex-stabilisierendes Metall-Salen-Basenpaar". Angew. Chem. Int. Ed. 117: 7370–7374. doi:10.1002/ange.200501589.

64. Buncel, E.; Boone, C.; Joly, H.; Kumar, R.; Norris, A. R. J. "Inorg. Biochem.". 1985'. 25: 61–73.

65. Ono, A.; Togashi, H. (2004). "Highly Selective Oligonucleotide-Based Sensor for Mercury(II) in Aqueous Solutions". Angew. Chem. 43: 4300–4302. doi:10.1002/anie.200454172.

66. Meggers, E.; Holland, P. L.; Tolman, W. B.; Romesberg, F. E.; Schultz, P. G. "J. Am. Chem. Soc.". 2000'. 122: 10714–10715. doi:10.1021/ja0025806.

67. Lee, J. S.; Skinner, R. J. S.; Latimer, L. J. P.; Reid, R. S. "Biochem. Cell Biol.". 1993'. 71: 162–168. doi:10.1139/o93-026.

68. Tanaka, K.; Tengeiji, A.; Kato, T.; Toyama, N.; Shionoya, M. (2003). "A Discrete Self-Assembled Metal Array in Artificial DNA". Science. 299: 1212–1213. doi:10.1126/science.1080587.

69. Malyshev, Denis A.; Dhami, Kirandeep; Quach, Henry T.; Lavergne, Thomas; Ordoukhanian, Phillip (24 July 2012). "Efficient and sequence-independent replication of DNA containing a third base pair establishes a functional six-letter genetic alphabet". Proceedings of the National Academy of Sciences of the United States of America. 109 (30): 12005–12010. Bibcode:2012PNAS..10912005M. doi:10.1073/pnas.1205176109. Retrieved 2014-05-11.

70. Callaway, Ewan (May 7, 2014). "Scientists Create First Living Organism With 'Artificial' DNA". Nature News. Huffington Post. Retrieved 8 May 2014.

71. Fikes, Bradley J. (May 8, 2014). "Life engineered with expanded genetic code". San Diego Union Tribune. Retrieved 8 May 2014.

72. Sample, Ian (May 7, 2014). "First life forms to pass on artificial DNA engineered by US scientists". The Guardian. Retrieved 8 May 2014.

73. Pollack, Andrew (May 7, 2014). "Scientists Add Letters to DNA's Alphabet, Raising Hope and Fear". New York Times. Retrieved 8 May 2014.

74. Hirao, I.; et al. (2002). "An unnatural base pair for incorporating amino acid analogs into proteins". Nat. Biotechnol. 20: 177–182. doi:10.1038/nbt0202-177.

75. Hirao, I.; et al. (2006). "An unnatural hydrophobic base pair system: site-specific incorporation of nucleotide analogs into DNA and RNA". Nat. Methods. 6: 729–735. doi:10.1038/nmeth915.

76. Kimoto, M.; et al. (2009). "An unnatural base pair system for efficient PCR amplification and functionalization of DNA molecules". Nucleic acids Res. 37: e14. doi:10.1093/nar/gkn956.

77. Yamashige, R.; et al. "Highly specific unnatural base pair systems as a third base pair for PCR amplification". Nucleic Acids Res. 40: 2793–2806. doi:10.1093/nar/gkr1068.

78. Kimoto, M.; et al. (2013). "Generation of high-affinity DNA aptamers using an expanded genetic alphabet". Nat. Biotechnol. 31: 453–457. doi:10.1038/nbt.2556.

79. Schmidt M. Xenobiology: a new form of life as the ultimate biosafety tool Bioessays Vol 32(4):322-331

80. Herdewijn, P; Marlière, P (Jun 2009). "Toward safe genetically modified organisms through the chemical diversification of nucleic acids". Chem Biodivers. 6 (6): 791–808. doi:10.1002/cbdv.200900083.

81. Shinkai, A.; Patel, P. H.; Loeb, L. A. (2001). J Biol. Chem. 276: 18836. Missing or empty |title= (help)

82. Rackham, O; Chin, JW (Aug 2005). "A network of orthogonal ribosome x mRNA pairs". Nat Chem Biol. 1 (3): 159–66. doi:10.1038/nchembio719.

CHAPTER 3

Carbohydrates Biomolecules

Chapter Contents

- **Introduction**
- **Structure**
- **Monosaccharides**
- **Nutrition**

Carbohydrates, or saccharides, are sugars and starches, which provide energy for humans and animals, and cellulose which make up many plant structures. "Carbs," as they are now commonly referred to, have become both a blessing and a curse, as the process of modern food production has changed the way they are consumed. There are two types of carbohydrates, simple, or monosaccharides and complex, or polysaccharides.

INTRODUCTION

A carbohydrate is a biological molecule consisting of carbon (C), hydrogen (H) and oxygen (O) atoms, usually with a hydrogen–oxygen atom ratio of 2:1 (as in water); in other words, with the empirical formula $C_m(H_2O)_n$ (where m could be different from n). Some exceptions exist; for example, deoxyribose, a sugar component of DNA, has the empirical formula $C_5H_{10}O_4$. Carbohydrates are technically hydrates of carbon; structurally it is more accurate to view them as polyhydroxy aldehydes and ketones.

The term is most common in biochemistry, where it is a synonym of **saccharide**, a group that includes sugars, starch, and cellulose. The saccharides are divided into four chemical groups: monosaccharides, disaccharides, oligosaccharides, and polysaccharides. In general, the monosaccharides and disaccharides, which are smaller (lower molecular weight) carbohydrates, are commonly referred to as sugars. The word saccharide comes from the Greek word *σάκχαρον* (sákkharon), meaning "sugar". While the scientific nomenclature of carbohydrates is complex, the names of the monosaccharides and disaccharides very often end in the suffix -ose. For example, grape sugar is the monosaccharide glucose, cane sugar is the disaccharide sucrose, and milk sugar is the disaccharide lactose (see illustration).

Carbohydrates perform numerous roles in living organisms. Polysaccharides serve for the storage of energy (e.g. starch and glycogen) and as structural components (e.g. cellulose in plants and chitin in arthropods). The 5-carbon monosaccharide ribose is an important component of coenzymes (e.g. ATP, FAD and NAD) and the backbone of the genetic molecule known as RNA. The related deoxyribose is a component of DNA. Saccharides and their derivatives include many other important biomolecules that play key roles in the immune system, fertilization, preventing pathogenesis, blood clotting, and development.

In food science and in many informal contexts, the term carbohydrate often means any food that is particularly rich in the complex carbohydrate starch (such as cereals, bread and pasta) or simple carbohydrates, such as sugar (found in candy, jams, and desserts).

Often in lists of nutritional information, such as the USDA National Nutrient Database, the term "carbohydrate" (or "carbohydrate by difference") is used for everything other than water, protein, fat, ash, and ethanol. This will include chemical compounds such as acetic or lactic acid, which are not normally considered carbohydrates. It also includes "dietary fiber" which is a carbohydrate but which does not contribute much in the way of food energy (calories), even though it is often included in the calculation of total food energy just as though it were a sugar.

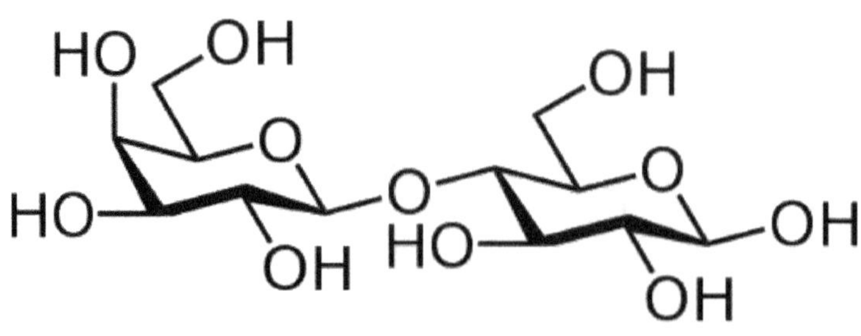

Lactose is a disaccharide found in milk. It consists of a molecule of D-galactose and a molecule of D-glucose bonded by beta-1-4 glycosidic linkage. It has a formula of $C_{12}H_{22}O_{11}$.

STRUCTURE

Formerly the name "carbohydrate" was used in chemistry for any compound with the formula $C_m(H_2O)_n$. Following this definition, some chemists considered formaldehyde (CH_2O) to be the simplest carbohydrate, while others claimed that title for glycolaldehyde. Today, the term is generally understood in the biochemistry sense, which excludes compounds with only one or two carbons and includes many biological carbohydrates which deviate from this formula. For example, while the above representative formulas would seem to capture the commonly known carbohydrates, ubiquitous and abundant carbohydrates often deviate from this. For example, carbohydrates often display chemical groups such as: N-acetyl (e.g. chitin), sulphate (e.g. glycosaminoglycans), carboxylic acid (e.g. sialic acid) and deoxy modifications (e.g. fucose and sialic acid).

Natural saccharides are generally built of simple carbohydrates called monosaccharides with general formula $(CH_2O)_n$ where n is three or more. A typical monosaccharide has the structure H–$(CHOH)_x$(C=O)–$(CHOH)_y$–H, that is, an aldehyde or ketone with many hydroxyl groups added, usually one on each carbon atom that is not part of the aldehyde or ketone functional group. Examples of monosaccharides are glucose, fructose, and glyceraldehydes. However, some biological substances commonly called "monosaccharides" do not conform to this formula (e.g. uronic acids and deoxy-sugars such as fucose) and there are many chemicals that do conform to this formula but are not considered to be monosaccharides (e.g. formaldehyde CH_2O and inositol $(CH_2O)_6$).

The open-chain form of a monosaccharide often coexists with a closed ring form where the aldehyde/ketone carbonyl group carbon (C=O) and hydroxyl group (–OH) react forming a hemiacetal with a new C–O–C bridge.

Monosaccharides can be linked together into what are called polysaccharides (or oligosaccharides) in a large variety of ways. Many carbohydrates contain one or more modified monosaccharide units that have had one or more groups replaced or removed. For example, deoxyribose, a component of DNA, is a modified version of ribose; chitin is composed of repeating units of N-acetyl glucosamine, a nitrogen-containing form of glucose.

Division

Carbohydrates are polyhydroxy aldehydes, ketones, alcohols, acids, their simple derivatives and their polymers having linkages of the acetal type. They may be classified according to their degree of polymerization and may be divided initially into three principal groups, namely sugars, oligosaccharides and polysaccharides

The major dietary carbohydrates

Class (DP*)	**Subgroup**	**Components**
Sugars (1–2)	Monosaccharides	Glucose, galactose, fructose, xylose
	Disaccharides	Sucrose, lactose, maltose, trehalose
	Polyols	Sorbitol, mannitol

Oligosaccharides (3–9)	Malto-oligosaccharides	Maltodextrins
	Other oligosaccharides	Raffinose, stachyose, fructo-oligosaccharides
Polysaccharides (>9)	Starch	Amylose, amylopectin, modified starches
	Non-starch polysaccharides	Cellulose, hemicellulose, pectins, hydrocolloids

DP * = Degree of polymerization

MONOSACCHARIDES

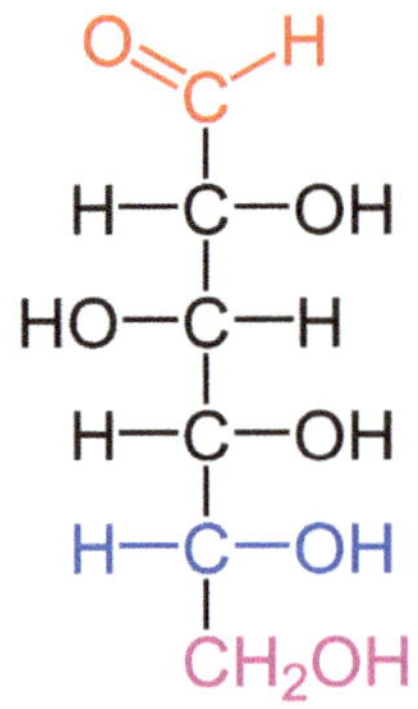

D-glucose is an aldohexose with the formula $(C{\cdot}H_2O)_6$. The red atoms highlight the aldehyde group and the blue atoms highlight the asymmetric center furthest from the aldehyde; because this -OH is on the right of the Fischer projection, this is a D sugar.

Monosaccharides are the simplest carbohydrates in that they cannot be hydrolyzed to smaller carbohydrates. They are aldehydes or ketones with two or more hydroxyl groups. The general chemical formula of an unmodified monosaccharide is $(C{\bullet}H_2O)_n$, literally a "carbon hydrate". Monosaccharides are important fuel molecules as well as building blocks for nucleic acids. The smallest monosaccharides, for which n=3, are dihydroxyacetone and D- and L-glyceraldehydes.

Classification of Monosaccharides

Monosaccharides are classified according to three different characteristics: the placement of its carbonyl group, the number of carbon atoms it contains, and its chiral handedness. If the carbonyl group is an aldehyde, the monosaccharide is an aldose; if the carbonyl group is a ketone, the monosaccharide is a ketose. Monosaccharides with three carbon atoms are called trioses, those with four are called tetroses, five are called pentoses, six are hexoses, and so on. These two systems of classification are often combined. For example, glucose is an aldohexose (a six-carbon aldehyde), ribose is an aldopentose (a five-carbon aldehyde), and fructose is a ketohexose (a six-carbon ketone).

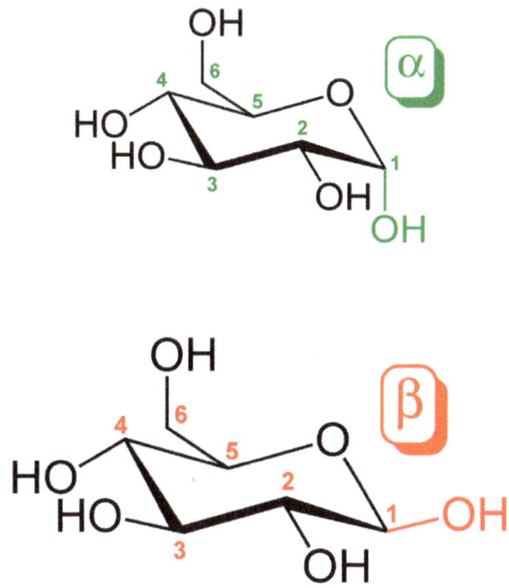

The α and β anomers of glucose. Note the position of the hydroxyl group (red or green) on the anomeric carbon relative to the CH_2OH group bound to carbon 5: they either have identical absolute configurations (R,R or S,S) (α), or opposite absolute configurations (R,S or S,R) (β).

Each carbon atom bearing a hydroxyl group (-OH), with the exception of the first and last carbons, are asymmetric, making them stereo centers with two possible configurations each (R or S). Because of this asymmetry, a number of isomers may exist for any given monosaccharide formula. Using Le Bel-van›t Hoff rule, the aldohexose D-glucose, for example, has the formula $(C{\cdot}H_2O)_6$, of which four of its six carbons atoms are stereogenic, making D-glucose one of 2^4=16 possible stereoisomers. In the case of glyceraldehydes, an aldotriose, there is one pair of possible stereoisomers, which are enantiomers and epimers. 1, 3-dihydroxyacetone, the ketose corresponding to the aldose glyceraldehydes, is a symmetric molecule with no stereo centers. The assignment of D or L is made according to the orientation of the asymmetric carbon furthest from the carbonyl group: in a standard Fischer projection if the hydroxyl group is on the right the molecule is a D sugar, otherwise it is an L sugar. The "D-" and "L-" prefixes should not be confused with "d-" or "l-", which indicate the direction that the sugar rotates plane polarized light. This usage of "d-" and "l-" is no longer followed in carbohydrate chemistry.

Structure and Nomenclature

With few exceptions (e.g., deoxyribose), monosaccharides have this chemical formula: $C_x(H_2O)_y$, where conventionally $x \geq 3$. Monosaccharides can be classified by the number x of carbon atoms they contain: triose (3) tetrose (4), pentose (5), hexose (6), heptose (7), and so on. The most important monosaccharide, glucose, is a hexose. Examples of heptoses include the ketoses mannoheptulose and sedoheptulose. Monosaccharides with eight or more carbons are rarely observed as they are quite unstable. In aqueous solutions monosaccharides exist as rings.

Linear-Chain Monosaccharides

Simple monosaccharides have a linear and unbranched carbon skeleton with one carbonyl (C=O) functional group, and one hydroxyl (OH) group on each of the remaining carbon atoms. Therefore, the molecular structure of a simple monosaccharide can be written as $H(CHOH)_n(C{=}O)(CHOH)_mH$, where n + 1 + m = x; so that its elemental formula is $C_xH_{2x}O_x$.

By convention, the carbon atoms are numbered from 1 to x along the backbone, starting from the end that is closest to the C=O group.

If the carbonyl is at position 1 (that is, n or m is zero), the molecule begins with a formyl group H(C=O)− and is technically an aldehyde. In that case, the compound is termed an aldose. Otherwise, the molecule has a keto group, a carbonyl −(C=O)− between two carbons; then it is formally a ketone, and is termed a ketose. Ketoses of biological interest usually have the carbonyl at position 2.

The various classifications above can be combined, resulting in names such as "aldohexose" and "ketotriose".

A more general nomenclature for open-chain monosaccharides combines a Greek prefix to indicate the number of carbons (tri-, tetr-, pent-, hex-, etc.) with the suffixes "-ose" for aldoses and "-ulose" for ketoses. In the latter case, if the carbonyl is not at position 2, its position is then indicated by a numeric infix. So, for example, $H(C{=}O)(CHOH)_4H$ is pentose, $H(CHOH)(C{=}O)(CHOH)_3H$ is pentulose, and $H(CHOH)_2(C{=}O)(CHOH)_2H$ is pent-3-ulose.

Open-Chain Stereoisomers

Two monosaccharides with equivalent molecular graphs (same chain length and same carbonyl position) may still be distinct stereoisomers, whose molecules differ in the three-dimensional arrangement of the bonds of certain atoms. This happens only if the molecule contains a stereogenic center, specifically a carbon atom that is chiral (connected to four distinct molecular sub-structures). Those four bonds can have any of two configurations in space distinguished by their handedness. In a simple open-chain monosaccharide, every carbon is chiral except the first and the last atoms of the chain, and (in ketoses) the carbon with the keto group.

For example, the triketose H(CHOH)(C=O)(CHOH)H (glycerone, dihydroxyacetone) has no stereogenic center, and therefore exists as a single stereoisomer. The other triose, the aldose $H(C{=}O)(CHOH)_2H$ (glyceraldehyde), has one chiral carbon — the central one, number 2 — which is bonded to groups −H, −OH, $-C(OH)H_2$, and −(C=O)H. Therefore, it exists as two stereoisomers whose molecules are mirror images of each other (like a left and a right glove). Monosaccharides with four or more carbons may contain multiple chiral carbons, so they typically have more than two stereoisomers. The number of distinct stereoisomers with the same diagram is bounded by 2^c, where c is the total number of chiral carbons.

The Fischer projection is a systematic way of drawing the skeletal formula of an acyclic monosaccharide so that the handedness of each chiral carbon is well specified.

Each stereoisomer of a simple open-chain monosaccharide can be identified by the positions (right or left) in the Fischer diagram of the chiral hydroxyls (the hydroxyls attached to the chiral carbons).

Most stereoisomers are themselves chiral (distinct from their mirror images). In the Fischer projection, two mirror-image isomers differ by having the positions of all chiral hydroxyls reversed right-to-left. Mirror-image isomers are chemically identical in non-chiral environments, but usually have very different biochemical properties and occurrences in nature.

While most stereoisomers can be arranged in pairs of mirror-image forms, there are some non-chiral stereoisomers that are identical to their mirror images, in spite of having chiral centers. This happens whenever the molecular graph is symmetrical, as in the 3-ketopentoses $H(CHOH)_2(CO)(CHOH)_2H$, and the two halves are mirror images of each other. In that case, mirroring is equivalent to a half-turn rotation. For this reason, there are only three distinct 3-ketopentose stereoisomers, even though the molecule has two chiral carbons.

Distinct stereoisomers that are not mirror-images of each other usually have different chemical properties, even in non-chiral environments. Therefore, each mirror pair and each non-chiral stereoisomer may be given a specific monosaccharide name. For example, there are 16 distinct aldohexose stereoisomers, but the name "glucose" means a specific pair of mirror-image aldohexoses. In the Fischer projection, one of the two glucose isomers has the hydroxyl at left on C3, and at right on C4 and C5; while the other isomer has the reversed pattern. These specific monosaccharide names have conventional three-letter abbreviations, like "Glu" for glucose and "Thr" for threose.

Generally, a monosaccharide with n asymmetrical carbons has 2^n stereoisomers. The number of open chain stereoisomers for an aldose monosaccharide is larger by one than that of a ketose monosaccharide of the same length. Every ketose will have $2^{(n-3)}$ stereoisomers where $n > 2$ is the number of carbons. Every aldose will have $2^{(n-2)}$ stereoisomers where $n > 2$ is the number of carbons. These are also referred to as epimers which have the different arrangement of −OH and −H groups at the asymmetric or chiral carbon atoms (this does not apply to those carbons having the carbonyl functional group).

Configuration of Monosaccharides

Like many chiral molecules, the two stereoisomers of glyceraldehyde will gradually rotate the polarization direction of linearly polarized light as it passes through it, even in solution. The two stereoisomers are identified with the prefixes D- and L-, according to the sense of rotation: D-glyceraldehyde is dextrorotatory (rotates the polarization axis clockwise), while L-glyceraldehyde is levorotatory (rotates it counterclockwise).

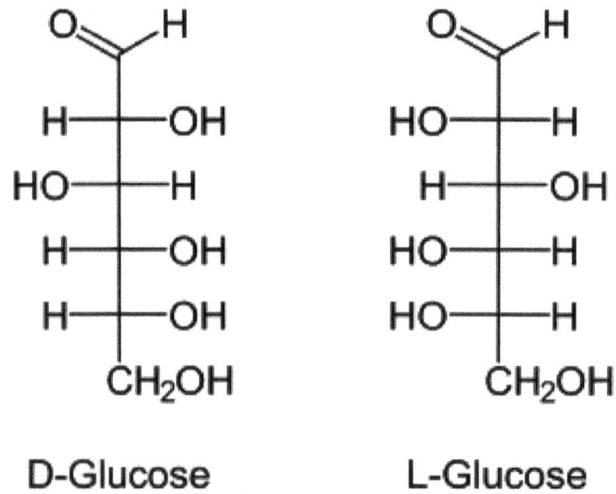

D- and L-glucose

The D- and L- prefixes are also used with other monosaccharides, to distinguish two particular stereoisomers that are mirror-images of each other. For this purpose, one considers the chiral carbon that is furthest removed from the C=O group. Its four bonds must connect to −H, −OH, −C(OH)H, and the rest of the molecule. If the molecule can be rotated in space so that the directions of those four groups match those of the analog groups in D-glyceraldehyde's C2, then the isomer receives the D- prefix. Otherwise, it receives the L- prefix.

In the Fischer projection, the D- and L- prefixes specifies the configuration at the carbon atom that is second from bottom: D- if the hydroxyl is on the right side, and L- if it is on the left side.

Note that the D- and L- prefixes do not indicate the direction of rotation of polarized light, which is a combined effect of the arrangement at all chiral centers. However, the two enantiomers will always rotate the light in opposite directions, by the same amount.

Cyclisation of Monosaccharides

A monosaccharide often switches from the acyclic (open-chain) form to a cyclic form, through a nucleophilic addition reaction between the carbonyl group and one of the hydroxyls of the same molecule. The reaction creates a ring of carbon atoms closed by one bridging oxygen atom. The resulting molecule has an hemiacetal or hemiketal group, depending on whether the linear form was an aldose or a ketose. The reaction is easily reversed, yielding the original open-chain form.

In these cyclic forms, the ring usually has **5** or **6** atoms. These forms are called furanoses and pyranoses, respectively — by analogy with furan and pyran, the simplest compounds with the same carbon-oxygen ring (although they lack the double bonds of these two molecules). For example, the aldohexose glucose may form a hemiacetal linkage between the hydroxyl on carbon 1 and the oxygen on carbon 4, yielding a molecule with a 5-membered ring, called glucofuranose. The same reaction can take place between carbons 1 and 5 to form a molecule with a 6-membered ring, called glucopyranose. Cyclic forms with a 7-atom ring (the same of oxepane), rarely encountered, are called heptoses.

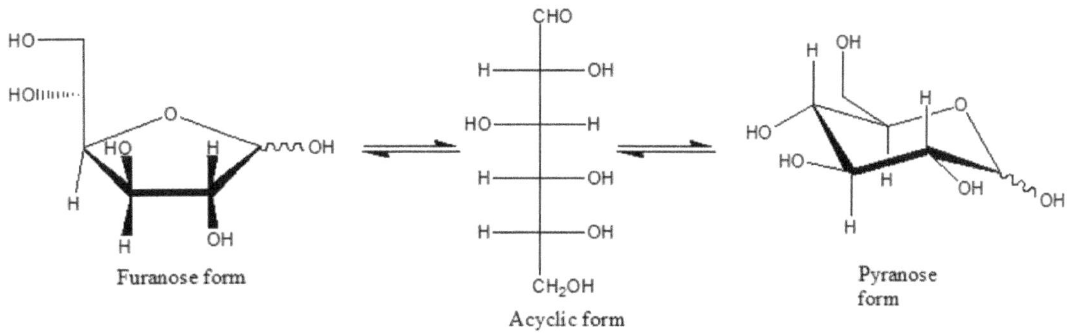

Conversion between the furanose, acyclic, and pyranose forms of D-glucose.

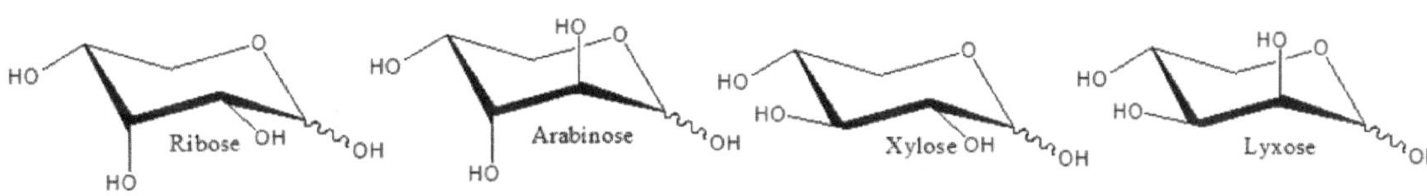

Pyranose forms of some pentose sugars.

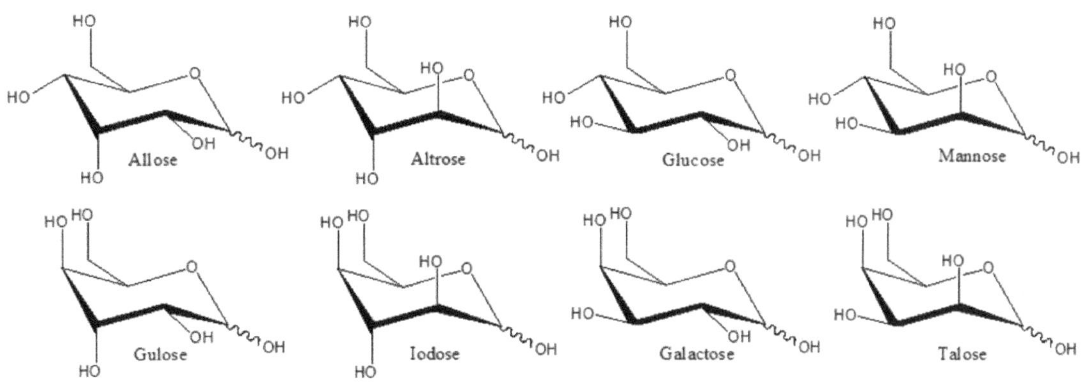

Pyranose forms of some hexose sugars.

For many monosaccharides (including glucose), the cyclic forms predominate, in the solid state and in solutions, and therefore the same name commonly is used for the open- and closed-chain isomers. Thus, for example, the term "glucose" may signify glucofuranose, glucopyranose, the open-chain form, or a mixture of the three.

Cyclization creates a new stereogenic center at the carbonyl-bearing carbon. The −OH group that replaces the carbonyl's oxygen may end up in two distinct positions relative to the ring's midplane. Thus each open-chain monosaccharide yields two cyclic isomers (anomers), denoted by the prefixes α- and β-. The molecule can change between these two forms by a process called mutarotation, that consists in a reversal of the ring-forming reaction followed by another ring formation.

Haworth Projection

The stereochemical structure of a cyclic monosaccharide can be represented in a Haworth projection. In this diagram, the α-isomer has the -OH of the anomeric carbon below the plane of the carbon atoms, and the β-isomer has the -OH of the anomeric carbon above the plane. Pyranoses typically adopt a chair conformation, similar to that of cyclohexane. In this conformation, the α-isomer has the -OH of the anomeric carbon in an axial position, whereas the β-isomer has the OH- of the anomeric carbon in equatorial position.

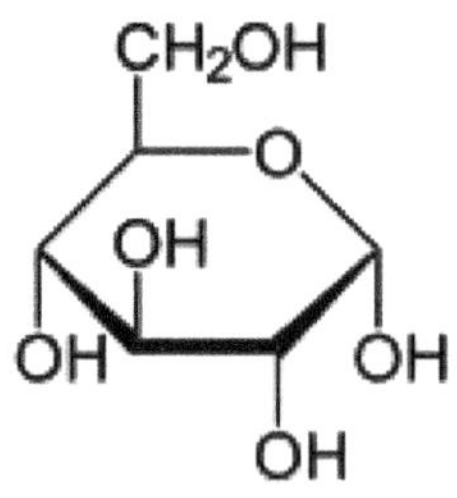

α-D-Glucopyranose

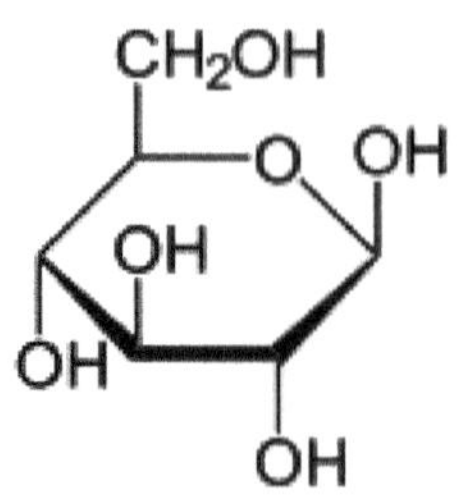

β-D-Glucopyranose

Derivatives

A large number of biologically important modified monosaccharides exist:

- Amino sugars such as:
  - Galactosamine
  - Glucosamine

  - Sialic acid
  - N-Acetylglucosamine
- Sulfosugars such as:
  - Sulfoquinovose

Ring-Straight Chain Isomerism

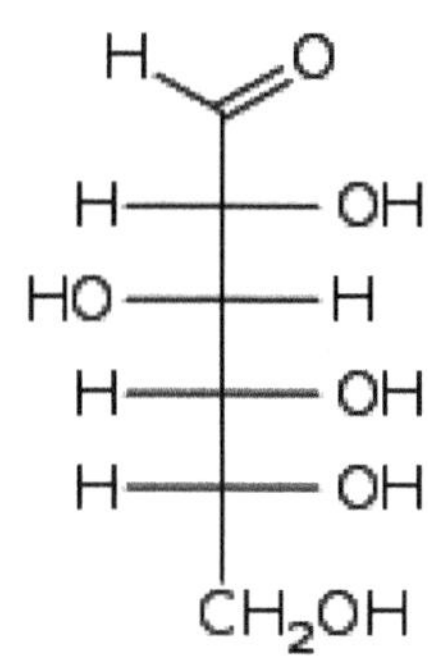

Glucose can exist in both a straight-chain and ring form.

The aldehyde or ketone group of a straight-chain monosaccharide will react reversibly with a hydroxyl group on a different carbon atom to form a hemiacetal or hemiketal, forming a heterocyclic ring with an oxygen bridge between two carbon atoms. Rings with five and six atoms are called furanose and pyranose forms, respectively, and exist in equilibrium with the straight-chain form.

During the conversion from straight-chain form to the cyclic form, the carbon atom containing the carbonyl oxygen, called the anomeric carbon, becomes a stereogenic center with two possible configurations: The oxygen atom may take a position either above or below the plane of the ring. The resulting possible pair of stereoisomers is called anomers. In the α anomer, the -OH substituent on the anomeric carbon rests on the opposite side (trans) of the ring from the CH_2OH side branch. The alternative form, in which the CH_2OH substituent and the anomeric hydroxyl are on the same side (cis) of the plane of the ring, is called the β anomer.

Use in Living Organisms

Monosaccharides are the major source of fuel for metabolism, being used both as an energy source (glucose being the most important in nature) and in biosynthesis. When monosaccharides are not immediately needed by many cells they are often converted to more space-efficient forms, often polysaccharides. In many animals, including humans, this storage form is glycogen, especially in liver and muscle cells. In plants, starch is used for the same purpose. The most abundant carbohydrate, cellulose, is a structural component of the cell wall of plants and many forms of algae. Ribose is a component of RNA.

Deoxyribose is a component of DNA. Lyxose is a component of lyxoflavin found in the human heart. Ribulose and xylulose occur in the pentose phosphate pathway. Galactose, a component of milk sugar lactose, is found in galactolipids in plant cell membranes and in glycoproteins in many tissues. Mannose occurs in human metabolism, especially in the glycosylation of certain proteins. Fructose, or fruit sugar, is found in many plants and in humans, it is metabolized in the liver, absorbed directly into the intestines during digestion, and found in semen. Trehalose, a major sugar of insects, is rapidly hydrolyzed into two glucose molecules to support continuous flight.

Disaccharides

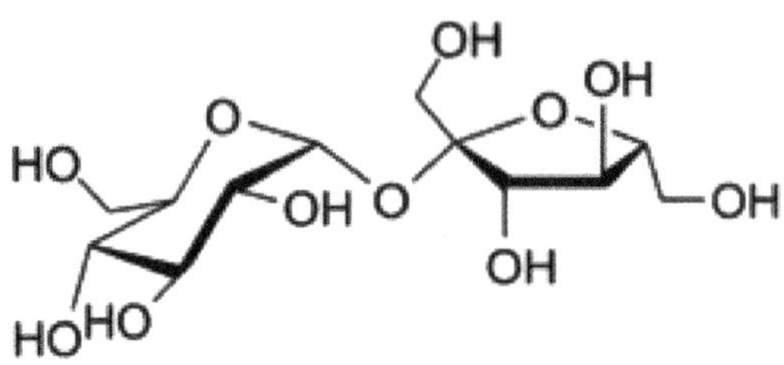

Sucrose, also known as table sugar, is a common disaccharide. It is composed of two monosaccharides: D-glucose (left) and D-fructose (right).

Two joined monosaccharides are called a disaccharide and these are the simplest polysaccharides. Examples include sucrose and lactose. They are composed of two monosaccharide units bound together by a covalent bond known as a glycosidic linkage formed via a dehydration reaction, resulting in the loss of a hydrogen atom from one monosaccharide and a hydroxyl group from the other. The formula of unmodified disaccharides is $C_{12}H_{22}O_{11}$. Although there are numerous kinds of disaccharides, a handful of disaccharides are particularly notable.

Sucrose, pictured to the right, is the most abundant disaccharide, and the main form in which carbohydrates are transported in plants. It is composed of one D-glucose molecule and one D-fructose molecule. The systematic name for sucrose, O-α-D-glucopyranosyl-(1→2)-D-fructofuranoside, indicates four things:

- Its monosaccharides: glucose and fructose
- Their ring types: glucose is a pyranose and fructose is a furanose
- How they are linked together: the oxygen on carbon number 1 (C1) of α-D-glucose is linked to the C2 of D-fructose.
- The -oside suffix indicates that the anomeric carbon of both monosaccharides participates in the glycosidic bond.

Lactose, a disaccharide composed of one D-galactose molecule and one D-glucose molecule, occurs naturally in mammalian milk. The systematic name for lactose is O-β-D-galactopyranosyl-(1→4)-D-glucopyranose. Other notable disaccharides include maltose (two D-glucoses linked α-1,4) and cellulobiose (two D-glucoses linked β-1,4). Disaccharides can be classified into two types: reducing and non-reducing disaccharides. If the functional group is present in bonding with another sugar unit, it is called a reducing disaccharide or biose.

NUTRITION

Carbohydrate consumed in food yields 3.87 calories of energy per gram for simple sugars, and 3.57 to 4.12 calories per gram for complex carbohydrate in most other foods. Relatively high levels of carbohydrate are associated with processed foods or refined foods made from plants, including sweets, cookies and candy, table sugar, honey, soft drinks, breads and crackers, jams and fruit products, pastas and breakfast cereals. Lower amounts of carbohydrate are usually associated with unrefined foods, including beans, tubers, rice, and unrefined fruit. Animal-based foods generally have the lowest carbohydrate levels, although milk does contain a high proportion of lactose.

Carbohydrates are a common source of energy in living organisms; however, no carbohydrate is an essential nutrient in humans. Humans are able to obtain all of their energy requirement from protein and fats, though the potential for some negative health effects of extreme carbohydrate restriction remains, as the issue has not been studied extensively yet. However, in the case of dietary fiber – indigestible carbohydrates which are not a source of energy – inadequate intake can lead to significant increases in mortality.

Grain products: rich sources of carbohydrates

Following a diet consisting of very low amounts of daily carbohydrate for several days will usually result in higher levels of blood ketone bodies than an isocaloric diet with similar protein content. This relatively high level of ketone bodies is commonly known as ketosis and is very often confused with the potentially fatal condition often seen in type 1 diabetics known as diabetic ketoacidosis. Somebody suffering ketoacidosis will have much higher levels of blood ketone bodies along with high blood sugar, dehydration and electrolyte imbalance.

Long-chain fatty acids cannot cross the blood–brain barrier, but the liver can break these down to produce ketones. However, the medium-chain fatty acids octanoic and heptanoic acids can cross the barrier and be used by the brain, which normally relies upon glucose for its energy. Gluconeogenesis allows humans to synthesize some glucose from specific amino acids: from the glycerol backbone in triglycerides and in some cases from fatty acids.

Organisms typically cannot metabolize all types of carbohydrate to yield energy. Glucose is a nearly universal and accessible source of energy. Many organisms also have the ability to metabolize other monosaccharides and disaccharides but glucose is often metabolized first. In Escherichia coli, for example, the lac operon will express enzymes for the digestion of lactose when it is present, but if both lactose and glucose are present the lac operon is repressed, resulting in the glucose being used first (see: Diauxie). Polysaccharides are also common sources of energy. Many organisms can easily break down starches into glucose; most organisms, however, cannot metabolize cellulose or other polysaccharides like chitin and arabinoxylans. These carbohydrate types can be metabolized by some bacteria and protists. Ruminants and termites, for example, use microorganisms to process cellulose. Even though these complex carbohydrates are not very digestible, they represent an important dietary element for humans, called dietary fiber. Fiber enhances digestion, among other benefits.

Based on the effects on risk of heart disease and obesity, the Institute of Medicine recommends that American and Canadian adults get between 45–65% of dietary energy from carbohydrates. The Food and Agriculture Organization and World Health Organization jointly recommend that national dietary guidelines set a goal of 55–75% of total energy from carbohydrates, but only 10% directly from sugars (their term for simple carbohydrates).

Classification

Nutritionists often refer to carbohydrates as either simple or complex. However, the exact distinction between these groups can be ambiguous. The term complex carbohydrate was first used in the U.S. Senate Select Committee on Nutrition and Human Needs publication Dietary Goals for the United States (1977) where it was intended to distinguish sugars from other carbohydrates (which were perceived to be nutritionally superior). However, the report put "fruit, vegetables and whole-grains" in the complex carbohydrate column, despite the fact that these may contain sugars as well as polysaccharides. This confusion persists as today some nutritionists use the term complex carbohydrate to refer to any sort of digestible saccharide present in a whole food, where fiber, vitamins and minerals are also found (as opposed to processed carbohydrates, which provide energy but few other nutrients). The standard usage, however, is to classify carbohydrates chemically:

simple if they are sugars (monosaccharides and disaccharides) and complex if they are polysaccharides (or oligosaccharides).

In any case, the simple vs. complex chemical distinction has little value for determining the nutritional quality of carbohydrates. Some simple carbohydrates (e.g. fructose) raise blood glucose slowly, while some complex carbohydrates (starches), especially if processed, raise blood sugar rapidly. The speed of digestion is determined by a variety of factors including which other nutrients are consumed with the carbohydrate, how the food is prepared, individual differences in metabolism, and the chemistry of the carbohydrate.

The USDA's Dietary Guidelines for Americans 2010 call for moderate- to high-carbohydrate consumption from a balanced diet that includes six one-ounce servings of grain foods each day, at least half from whole grain sources and the rest from enriched.

The glycemic index (GI) and glycemic load concepts have been developed to characterize food behavior during human digestion. They rank carbohydrate-rich foods based on the rapidity and magnitude of their effect on blood glucose levels. Glycemic index is a measure of how quickly food glucose is absorbed, while glycemic load is a measure of the total absorbable glucose in foods. The insulin index is a similar, more recent classification method that ranks foods based on their effects on blood insulin levels, which are caused by glucose (or starch) and some amino acids in food.

Metabolism

Carbohydrate metabolism denotes the various biochemical processes responsible for the formation, breakdown and interconversion of carbohydrates in living organisms.

The most important carbohydrate is glucose, a simple sugar (monosaccharide) that is metabolized by nearly all known organisms. Glucose and other carbohydrates are part of a wide variety of metabolic pathways across species: plants synthesize carbohydrates from carbon dioxide and water by photosynthesis storing the absorbed energy internally, often in the form of starch or lipids. Plant components are consumed by animals and fungi, and used as fuel for cellular respiration. Oxidation of one gram of carbohydrate yields approximately 4 kcal of energy, while the oxidation of one gram of lipids yields about 9 kcal. Energy obtained from metabolism (e.g., oxidation of glucose) is usually stored temporarily within cells in the form of ATP. Organisms capable of aerobic respiration metabolize glucose and oxygen to release energy with carbon dioxide and water as byproducts.

Catabolism

Catabolism is the metabolic reaction which cells undergo to extract energy. There are two major metabolic pathways of monosaccharide catabolism: glycolysis and the citric acid cycle.

In glycolysis, oligo/polysaccharides are cleaved first to smaller monosaccharides by enzymes called glycoside hydrolases. The monosaccharide units can then enter into monosaccharide catabolism. In some cases, as with humans, not all carbohydrate types are usable as the digestive and metabolic enzymes necessary are not present.

Carbohydrate Chemistry

Carbohydrate chemistry is a large and economically important branch of organic chemistry. Some of the main organic reactions that involve carbohydrates are:

- Carbohydrate acetalisation
- Cyanohydrin reaction
- Lobry-de Bruyn-van Ekenstein transformation
- Amadori rearrangement
- Nef reaction
- Wohl degradation
- Koenigs–Knorr reaction
- Carbohydrate digestion

REFERENCES

1. Western Kentucky University (May 29, 2013). "WKU BIO 113 Carbohydrates". wku.edu.
2. Eldra Pearl Solomon; Linda R. Berg; Diana W. Martin (2004). Biology. Cengage Learning. p. 52. ISBN 978-0534278281 – via google.books.com.
3. National Institute of Standards and Technology (2011). "Material Measurement Library D-erythro-Pentose, 2-deoxy-". nist.gov.
4. Long Island University (May 29, 2013). "The Chemistry of Carbohydrates" (PDF). brooklyn.liu.edu.
5. Purdue University (May 29, 2013). "Carbohydrates: The Monosaccharides". purdue.edu.
6. Flitsch, Sabine L.; Ulijn, Rein V (2003). "Sugars tied to the spot". Nature. 421 (6920): 219–20. doi:10.1038/421219a. PMID 12529622.
7. Maton, Anthea; Jean Hopkins; Charles William McLaughlin; Susan Johnson; Maryanna Quon Warner; David LaHart; Jill D. Wright (1993). Human Biology and Health. Englewood Cliffs, New Jersey, USA: Prentice Hall. pp. 52–59. ISBN 0-13-981176-1.
8. USDA National Nutrient Database, 2015, p. 13
9. USDA National Nutrient Database, 2015, p. 14
10. John Merle Coulter, Charler Reid Barnes, Henry Chandler Cowles (1930), A Textbook of Botany for Colleges and Universities"
11. Carl A. Burtis, Edward R. Ashwood, Norbert W. Tietz (2000), Tietz fundamentals of clinical chemistry
12. Matthews, C. E.; K. E. Van Holde; K. G. Ahern (1999) Biochemistry. 3rd edition. Benjamin Cummings. ISBN 0-8053-3066-6
13. "Carbohydrates in human nutrition - Chapter 1 - The role of carbohydrates

in nutrition". Food and Agriculture Organization of the United Nations. FAO.

14. Carolyn R Bertozzi. "Structural Basis of Glycan Diversity - Essentials of Glycobiology - NCBI Bookshelf". nih.gov.
15. Campbell, Neil A.; Brad Williamson; Robin J. Heyden (2006). Biology: Exploring Life. Boston, Massachusetts: Pearson Prentice Hall. ISBN 0-13-250882-6.
16. Pigman, Ward; Horton, D. (1972). "Chapter 1: Stereochemistry of the Monosaccharides". In Pigman and Horton. The Carbohydrates: Chemistry and Biochemistry Vol 1A (2nd ed.). San Diego: Academic Press. pp. 1–67.
17. Pigman, Ward; Anet, E.F.L.J. (1972). "Chapter 4: Mutarotations and Actions of Acids and Bases". In Pigman and Horton. The Carbohydrates: Chemistry and Biochemistry Vol 1A (2nd ed.). San Diego: Academic Press. pp. 165–194.
18. "lyxoflavin". Merriam-Webster.
19. "Show Foods". usda.gov.
20. "CALCULATION OF THE ENERGY CONTENT OF FOODS - ENERGY CONVERSION FACTORS". fao.org.
21. "Carbohydrate reference list" (PDF). www.diabetes.org.uk. Retrieved October 30, 2016.
22. ^ Jump up to:a b Westman, EC (2002). "Is dietary carbohydrate essential for human nutrition?". The American Journal of Clinical Nutrition. 75 (5): 951–3; author reply 953–4. PMID 11976176.
23. Park, Y; Subar, AF; Hollenbeck, A; Schatzkin, A (2011). "Dietary fiber intake and mortality in the NIH-AARP diet and health study". Archives of Internal Medicine. 171 (12): 1061–8. doi:10.1001/archinternmed.2011.18. PMC 3513325 . PMID 21321288.
24. "Ketogenic low-carbohydrate diets have no metabolic advantage over nonketogenic low-carbohydrate diets"(PDF). nutrition.org.
25. Douglas Ebert. "Energy Contribution of Octanoate to Intact Rat Brain Metabolism Measured by 13C Nuclear Magnetic Resonance Spectroscopy". jneurosci.org.
26. "Journal of Cerebral Blood Flow & Metabolism - Abstract of article: Heptanoate as a neural fuel: energetic and neurotransmitter precursors in normal and glucose transporter I-deficient (G1D) brain". nature.com.
27. "Integration of Metabolism". medbio.info.
28. Pichon, L; Huneau, JF; Fromentin, G; Tomé, D (2006). "A high-protein, high-fat, carbohydrate-free diet reduces energy intake, hepatic lipogenesis, and adiposity in rats". The Journal of Nutrition. 136 (5): 1256–60. PMID 16614413.
29. Tighe, P; Duthie, G; Vaughan, N; Brittenden, J; Simpson, WG; Duthie, S; Mutch, W; Wahle, K; Horgan, G; Thies, F. (2010). "Effect of increased consumption of whole-grain foods on blood pressure and other cardiovascular risk markers in healthy middle-aged persons: a randomized, controlled trial". The American Journal of Clinical Nutrition. 92 (4): 733–40. doi:10.3945/ajcn.2010.29417. PMID 20685951.

30. Food and Nutrition Board (2002/2005). Dietary Reference Intakes for Energy, Carbohydrate, Fiber, Fat, Fatty Acids, Cholesterol, Protein and Amino Acids. Washington, D.C.: The National Academies Press. Page 769. ISBN 0-309-08537-3.
31. Joint WHO/FAO expert consultation (2003). (PDF). Geneva: World Health Organization. pp. 55–56. ISBN 92-4-120916-X.
32. Joint WHO/FAO expert consultation (1998), Carbohydrates in human nutrition, chapter 1. ISBN 92-5-104114-8.
33. ^ Jump up to:a b "Carbohydrates". The Nutrition Source. Harvard School of Public Health. Retrieved April 3, 2013.
34. Jenkins, David; Alexandra L. Jenkins; Thomas M.S. Woleve; Lilian H. Thompson; A. Venkat Rao (February 1986). "Simple and Complex Carbohydrates". Nutrition Reviews. 44 (2): 44–49. doi:10.1111/j.1753-4887.1986.tb07585.x.
35. DHHS and USDA, Dietary Guidelines for Americans 2010.
36. Energetics of Cellular Respiration (Glucose Metabolism)

CHAPTER 4

Membrane Lipids

Chapter Contents

- **Introduction**
- **Biological Roles**
- **Major Classes**

Membrane lipids are integral for active transport across the membrane, many types of enzyme activity and membrane formation. Lipids are a group of compounds that include fats and oils and are insoluble in water, but soluble in alcohol. In other words, lipids do not dissolve in water, which is clearly shown when you add oil to water – they mix, but remain separate. There are many different types of lipids, but the major classes of membrane lipids are phospholipids, glycolipids, sphingolipids and cholesterol.

INTRODUCTION

Membrane lipids are a group of compounds (structurally similar to fats and oils) which form the double-layered surface of all cells (lipid bilayer). The three major classes of membrane lipids are phospholipids, glycolipids, and cholesterol. Lipids are amphiphilic: they have one end that is soluble in water ('polar') and an ending that is soluble in fat ('nonpolar'). By forming a double layer with the polar ends pointing outwards and the nonpolar ends pointing inwards membrane lipids can form a 'lipid bilayer' which keeps the watery interior of the cell separate from the watery exterior. The arrangements of lipids and various proteins, acting as receptors and channel pores in the membrane, control the entry and exit of other molecules and ions as part of the cell's metabolism. In order to perform physiological functions, membrane proteins are facilitated to rotate and diffuse laterally in two dimensional expanse of lipid bilayer by the presence of a shell of lipids closely attached to protein surface, called annular lipid shell.

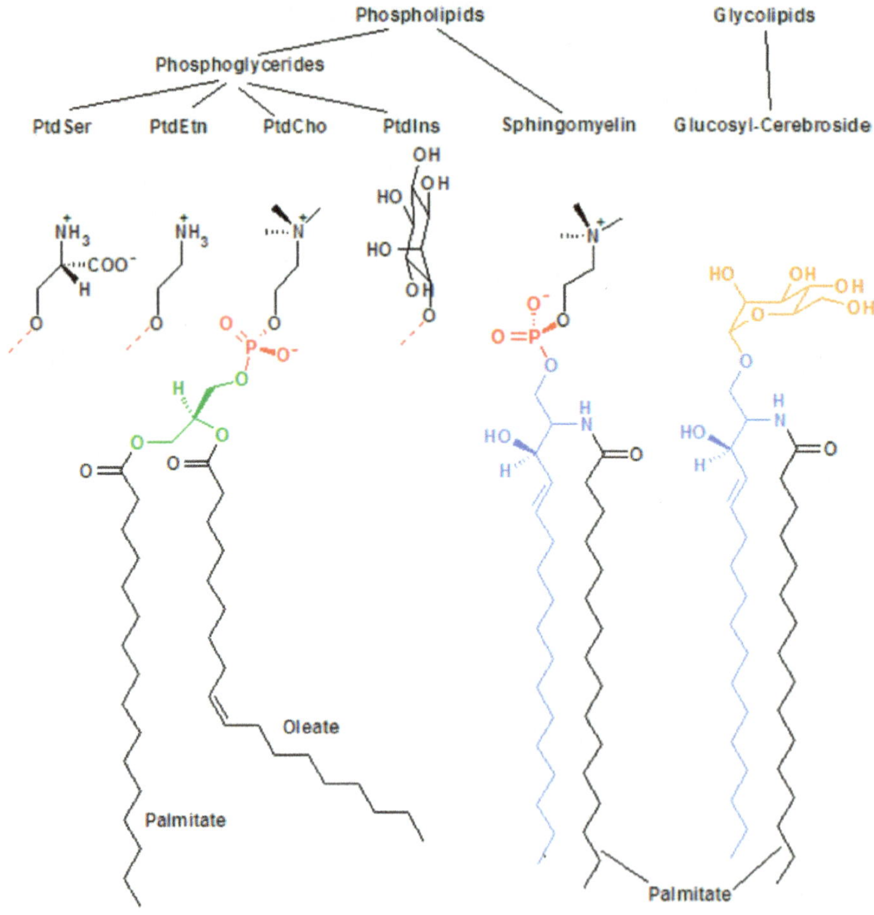

The major membrane lipids PtdCho - Phosphatidylcholine; PtdEtn - Phosphatidylethanolamine; PtdIns - Phosphatidylinositol; PtdSer - Phosphatidylserine.

BIOLOGICAL ROLES

The bilayer formed by membrane lipids serves as a containment unit of a living cell. Membrane lipids also form a matrix in which membrane proteins reside. Historically lipids were thought to merely serve a structural role. Functional roles of lipids are in fact many: They serve as regulatory agents in cell growth and adhesion. They participate in the biosynthesis of other biomolecules. They can serve to increase enzymatic activities of enzymes.

Non-bilayer forming lipid like monogalactosyl diglyceride(MGDG) predominates the bulk lipids in thylakoid membranes, which when hydrated alone, forms reverse hexagonal cylindrical phase. However, in combination with other lipids and carotenoids/chlorophylls of thylakoid membranes, they too conform together as lipid bilayers.

Membrane Lipid Language

The membrane metabolites of polyunsaturated fatty acids (PUFAs) have an essential role in intercellular biochemical communications. Crawford (2010) in his chapter *Long-chain polyunsaturated fatty acids in human brain evolution* reported, with regard to the *language of lipids,* that the importance of the increased complexity of these lipids was brought about by aerobic metabolism, whereby the simple language of prokaryotes, with only a few words, was developed into a vocabulary of over 1,000 words of eukaryote cells.

About 500 million years ago, some nervous cells and some gut cells of vertebrates, migrated and specialized in a more complex nerrvous system: the *brain*, and in uptake and storage of iodocompounds: the follicular *thyroid*. In the PUFAs, the presence of a double bond between two carbons (or carbon-carbon double bond) provides them with the possibility of changing their molecular structure through enzymes such as phospholipases, cyclooxygenases and lipoxygenases, etc. The resulting substances, called *eicosanoids*: prostaglandins (PG), leukotrienes (LT), lipoxins and tromboxane (TX); and docosanoids: resolvins, protectins, and maresins, are powerful lipid mediators that produce specific actions in the organism; they organize inflammation, hemodynamic, immune response and the repair of tissue.

Many PUFAs cannot be synthesized by animal organisms and are considered *essential*, and therefore should be incorporated into diets. These are: linoleic acid (C18:2 n-6), omega-6 and alpha-linolenic (C18:3 n-3) omega-3, arachidonic acid (AA) - omega - 6 (C20: 4n-6), and docosahexaenoic acid (DHA) - omega -3 (C22:6n-3). These PUFAs are incorporated into the phospholipidic membrane of all the cells of an organism. In parallel, ectodermic cells, differentiated into neuronal cells, became the primitive nervous system and brain. Both these cells synthesized iodolipids, as novel *words* of the chemical *lipid language* developed among cell membranes during the evolution of life. These biochemical signals among cells, since contact and modification of membranes in multicellular organisms formed the bases of adaptation to terrestrial environments, and their alterations are important in the mechanism of apoptosis, carcinogenesis and degenerative diseases, as well as for understand some problems discussed regarding human evolution (as Aquatic ape hypothesis) .

MAJOR CLASSES

Phospholipids

Phospholipids and glycolipids consist of two long, nonpolar (hydrophobic) hydrocarbon chains linked to a hydrophilic head group.

The heads of phospholipids are **phosphorylated** and they consist of either:

- **Glycerol** (and hence the name **phosphoglycerides** given to this group of lipids), or

- **Sphingosine** (e.g. sphingomyelin and ceramide).

Glycolipids

The heads of glycolipids (glyco- stands for sugar) contain a sphingosine with **one or several sugar units** attached to it. The hydrophobic chains belong either to:

- two fatty acids (FA) - in the case of the phosphoglycerides, or
- one FA and the hydrocarbon tail of sphingosine - in the case of sphingomyelin and the glycolipids.

Galactolipids- monogalactosyl diglyceride(MGDG) and digalactosyl diglycreride(DGDG) form the predominant lipids in higher plant chloroplast thylakoid membranes; liposomal structures formed by total lipid extract of thylakoid membranes have been found sensitive to sucrose as it turns bilayers into micellar structures.

Fatty acids

The fatty acids in phospho- and glycolipids usually contain an even number, typically between 14 and 24, of carbon atoms, with 16- and 18-carbon being the most common. FAs may be saturated or unsaturated, with the configuration of the double bonds nearly always *cis*. The length and the degree of unsaturation of FAs chains have a profound effect on membranes' fluidity. Plant *thylakoid membranes* maintain high fluidity, even at relatively cold environmental temperatures, due the abundance of 18-carbon fatty acyl chains with three double bonds, *linolenic acid*, as has been revealed by 13-C NMR studies.

Phosphoglycerides

In phosphoglycerides, the hydroxyl groups at C-1 and C-2 of glycerol are esterified to the carboxyl groups of the FAs. The C-3 hydroxyl group is esterified to phosphoric acid. The resulting compound, called phosphatidate, is the simplest phosphoglycerate. Only small amounts of phosphatidate are present in membranes. However, it is a key intermediate in the biosynthesis of the other phosphoglycerides.

Sphingolipids

Sphingosine is an amino alcohol that contains a long, unsaturated hydrocarbon chain. In sphingomyelin and glycolipids, the amino group of sphingosine is linked to FAs by an amide bond. In sphingomyelin the primary hydroxyl group of sphingosine is esterified to phosphoryl choline.

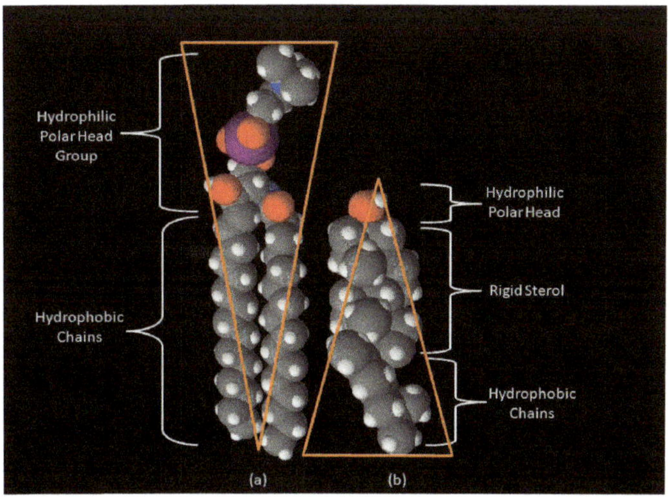

Space-filling models of (a) sphingomyelin and (b) cholesterol.

In glycolipids, the sugar component is attached to this group. The simplest glycolipid is cerebroside, in which there is only one sugar residue, either Glc or Gal. More complex glycolipids, such as gangliosides, contain a branched chain of as many as seven sugar residues.

Sterols

The best known sterol is cholesterol, which is found in humans. Cholesterol also occurs naturally in other eukaryote cell membranes. Sterols have a hydrophobic four-membered fused ring rigid structure, and a small polar head group.

Cholesterol is bio-synthesised from mevalonate via a squalene cyclisation of terpenoids. Cell membranes require high levels of cholesterol - typically an average of 20% cholesterol in the whole membrane, increasing locally in raft areas up to 50% cholesterol (- % is molecular ratio). It associates preferentially with sphingolipids (see diagram) in cholesterol-rich lipid rafts areas of the membranes in eukaryotic cells. Formation of lipid rafts promotes aggregation of peripheral and transmembrane proteins including docking of SNARE and VAMP proteins Phytosterols, such as sitosterol and stigmasterol, and Hopanoids serve a similar function in plants and prokaryotes.

REFERENCES

1. R. B. Gennis. Biomembranes - Molecular Structure and Function. Springer-Verlag, New York (1989).
2. YashRoy R.C. (1990) Lamellar dispersion and phase separation of chloroplast membrane lipids by negative staining electron microscopy. Journal of Biosciences, vol. 15(2), pp. 93-98.https://www.researchgate.net/publication/230820037_Lamellar_dispersion_and_phase_separation_of_chloroplast_membrane_lipids_

by_negative_staining_electron_microscopy?ev=prf_pub

3. Venturi, S.; Donati, F.M.; Venturi, A.; Venturi, M. (2000). "Environmental Iodine Deficiency: A Challenge to the Evolution of Terrestrial Life?". Thyroid. 10 (8): 727–9. doi:10.1089/10507250050137851. PMID 11014322.
4. Crawford, M. A.; Bloom, M.; Broadhurst, C. L.; Schmidt, W. F.; Cunnane, S. C.; Galli, C.; Gehbremeskel, K.; Linseisen, F.; Lloyd-Smith, J.; Parkington, J. (1999). "Evidence for the unique function of docosahexaenoic acid during the evolution of the modern hominid brain". Lipids. 34 Suppl: S39–S47. doi:10.1007/BF02562227. PMID 10419087.
5. Venturi, S; Bégin ME (2010). "Thyroid Hormone, Iodine and Human Brain Evolution". In Cunnane S; Stewart K. Environmental Influences on Human Brain Evolution. John Wiley & Sons. pp. 105–124. ISBN 978-0-470-45268-4.
6. Crawford MA (2010). "Long-Chain Polyunsaturated Fatty Acids in Human Brain Evolution". In Cunnane S; Stewart K. Environmental Influences on Human Brain Evolution. John Wiley & Sons. pp. 13–32. ISBN 978-0-470-45268-4.
7. Venturi, Sebastiano (2014). "Iodine, PUFAs and Iodolipids in Health and Disease: An Evolutionary Perspective". Human Evolution-. 29 (1-3): 185–205. ISSN 0393-9375.
8. YashRoy R.C. (1994) Destabilisation of lamellar dispersion of thylakoid membrane lipids by sucrose. Biochimica et Biophysica Acta, vol. 1212, pp. 129-133. https://www.researchgate.net/publication/15042978_Destabilisation_of_lamellar_dispersion_of_thylakoid_membrane_lipids_by_sucrose?ev=prf_pub
9. YashRoy R.C. (1987) 13-C NMR studies of lipid fatty acyl chains of chloroplast membranes. Indian Journal of Biochemistry and Biophysics, vol. 24(6), pp. 177-178.https://www.researchgate.net/publication/230822408_13-C_NMR_studies_of_lipid_fatty_acyl_chains_of_chloroplast_membranes?ev=prf_pub
10. de Meyer F, Smit B. Effect of cholesterol on the structure of a phospholipid bilayer. Proc Natl Acad Sci U S A 2009; 106: 3654-8.
11. Chen, Heshun; Born, Ella; Mathur, Satya N.; Field, F. Jeffrey (December 1, 1993). "Cholesterol and sphingomyelin syntheses are regulated independently in cultured human intestinal cells, CaCo-2: role of membrane cholesterol and sphingomyelin content." (PDF). Journal of Lipid Research. American Society for Biochemistry and Molecular Biology. 34 (12): 2159–67. ISSN 0022-2275.
12. Lang T, Bruns D, Wenzel D, Riedel D, Holroyd P, Thiele C, Jahn R. SNAREs are concentrated in cholesterol-dependent clusters that define docking and fusion sites for exocytosis EMBO J 2001;20:2202-13.

CHAPTER 5

Enzyme Kinetics

Chapter Contents

Enzyme kinetics is the study of the rates of enzyme-catalysed reactions.. The study of the rates of enzyme-catalysed reactions. Rates of reaction are usually measured by using the purified enzyme in vitro with the substrate and then observing the formation of the product or disappearance of the substrate. As the concentration of the substrate is increased the rate of reaction increases proportionally up to a certain point, after which any further increase in substrate concentration no longer increases the reaction rate (see Michaelis–Menten curve). At this point, all active sites of the enzyme are saturated with substrate; any further increase in the rate of reaction will occur only if more enzyme is added. Reaction rates are also affected by the presence of inhibitors (see inhibition), temperature, and pH.

INTRODUCTION

Enzyme kinetics is the study of the chemical reactions that are catalysed by enzymes. In enzyme kinetics, the reaction rate is measured and the effects of varying the conditions of the reaction are investigated. Studying an enzyme's kinetics in this way can reveal the catalytic mechanism of this enzyme, its role in metabolism, how its activity is controlled, and how a drug or an agonist might inhibit the enzyme. Enzymes are usually protein molecules that manipulate other molecules — the enzymes' substrates. These target molecules bind to an enzyme's active site and are transformed into products through a series of steps known as the enzymatic mechanism

$$E + S \rightleftarrows ES \rightleftarrows ES^* \rightleftarrows EP \rightleftarrows E + P$$

These mechanisms can be divided into single-substrate and multiple-substrate mechanisms. Kinetic studies on enzymes that only bind one substrate, such as triosephosphate isomerase, aim to measure the affinity with which the enzyme binds this substrate and the turnover rate. Some other examples of enzymes are phosphofructokinase and hexokinase, both of which are important for cellular respiration (glycolysis).

When enzymes bind multiple substrates, such as dihydrofolate reductase (shown right), enzyme kinetics can also show the sequence in which these substrates bind and the sequence in which products are released. An example of enzymes that bind a single substrate and release multiple products are proteases, which cleave one protein substrate into two polypeptide products. Others join two substrates together, such as DNA polymerase linking a nucleotide to DNA. Although these mechanisms are often a complex series of steps, there is typically one *rate-determining step* that determines the overall kinetics. This rate-determining step may be a chemical reaction or a conformational change of the enzyme or substrates, such as those involved in the release of product(s) from the enzyme.

Dihydrofolate reductase from E. coli with its two substrates dihydrofolate (right) and NADPH (left), bound in the active site. The protein is shown as a ribbon diagram, with alpha helices in red, beta sheets in yellow and loops in blue. Generated from 7DFR.

Knowledge of the enzyme's structure is helpful in interpreting kinetic data. For example, the structure can suggest how substrates and products bind during catalysis; what changes occur during the reaction; and even the role of particular amino acid residues in the mechanism. Some enzymes change shape significantly during the mechanism; in such cases, it is helpful to determine the enzyme structure with and without bound substrate analogues that do not undergo the enzymatic reaction.

Not all biological catalysts are protein enzymes; RNA-based catalysts such as ribozymes and ribosomes are essential to many cellular functions, such as RNA splicing and translation. The main difference between ribozymes and enzymes is that RNA catalysts are composed of nucleotides, whereas enzymes are composed of amino acids. Ribozymes also perform a more limited set of reactions, although their reaction mechanisms and kinetics can be analysed and classified by the same methods.

GENERAL PRINCIPLES

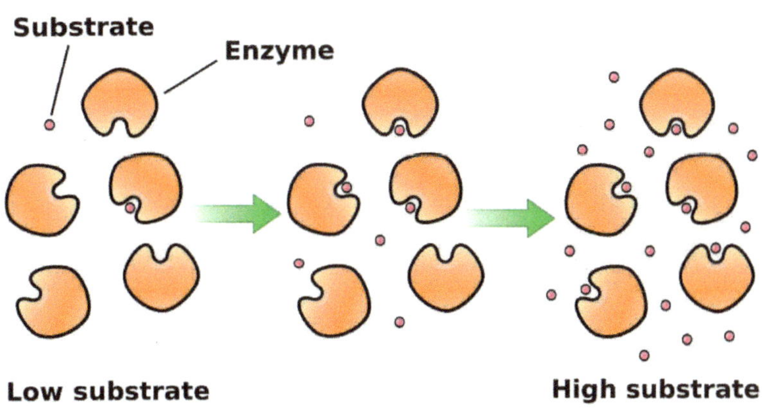

As larger amounts of substrate are added to a reaction, the available enzyme binding sites become filled to the limit of V_{max}. Beyond this limit the enzyme is saturated with substrate and the reaction rate ceases to increase.

The reaction catalysed by an enzyme uses exactly the same reactants and produces exactly the same products as the uncatalysed reaction. Like other catalysts, enzymes do not alter the position of equilibrium between substrates and products. However, unlike uncatalysed chemical reactions, enzyme-catalysed reactions display saturation kinetics. For a given enzyme concentration and for relatively low substrate concentrations, the reaction rate increases linearly with substrate concentration; the enzyme molecules are largely free to catalyse the reaction, and increasing substrate concentration means an increasing rate at which the enzyme and substrate molecules encounter one another. However, at relatively high substrate concentrations, the reaction rate asymptotically approaches the theoretical maximum; the enzyme active sites are almost all occupied and the reaction rate is determined by the intrinsic turnover rate of the enzyme. The substrate concentration midway between these two limiting cases is denoted by K_M.

The two most important kinetic properties of an enzyme are how quickly the enzyme becomes saturated with a particular substrate, and the maximum rate it can achieve. Knowing these properties suggests what an enzyme might do in the cell and can show how the enzyme will respond to changes in these conditions.

Enzyme Assays

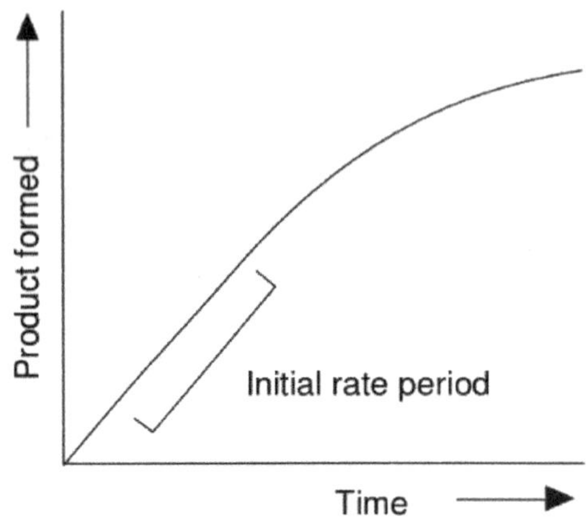

Progress curve for an enzyme reaction. The slope in the initial rate period is the initial rate of reaction v. The Michaelis–Menten equation describes how this slope varies with the concentration of substrate.

Enzyme assays are laboratory procedures that measure the rate of enzyme reactions. Because enzymes are not consumed by the reactions they catalyse, enzyme assays usually follow changes in the concentration of either substrates or products to measure the rate of reaction. There are many methods of measurement. Spectrophotometric assays observe change in the absorbance of light between products and reactants; radiometric assays involve the incorporation or release of radioactivity to measure the amount of product made over time. Spectrophotometric assays are most convenient since they allow the rate of the reaction to be measured continuously. Although radiometric assays require the removal and counting of samples (i.e., they are discontinuous assays) they are usually extremely sensitive and can measure very low levels of enzyme activity. An analogous approach is to use mass spectrometry to monitor the incorporation or release of stable isotopes as substrate is converted into product.

The most sensitive enzyme assays use lasers focused through a microscope to observe changes in single enzyme molecules as they catalyse their reactions. These measurements either use changes in the fluorescence of cofactors during an enzyme's reaction mechanism, or of fluorescent dyes added onto specific sites of the protein to report movements that occur during catalysis. These studies are providing a new view of the kinetics and dynamics of single enzymes, as opposed to traditional enzyme kinetics, which observes the average behaviour of populations of millions of enzyme molecules.

An example progress curve for an enzyme assay is shown above. The enzyme produces product at an initial rate that is approximately linear for a short period after the start of the reaction. As the reaction proceeds and substrate is consumed, the rate continuously slows (so long as substrate is not still at saturating levels). To measure the initial (and maximal) rate, enzyme assays are typically carried out while the reaction has progressed only a few percent towards total completion. The length of the initial rate period depends on the assay conditions and can range from milliseconds to hours. However, equipment for rapidly mixing liquids allows fast kinetic measurements on initial rates of less than one second. These very rapid assays are essential for measuring pre-steady-state kinetics, which are discussed below.

Most enzyme kinetics studies concentrate on this initial, approximately linear part of enzyme reactions. However, it is also possible to measure the complete reaction curve and fit this data to a non-linear rate equation. This way of measuring enzyme reactions is called progress-curve analysis. This approach is useful as an alternative to rapid kinetics when the initial rate is too fast to measure accurately.

SINGLE-SUBSTRATE REACTIONS

Enzymes with single-substrate mechanisms include isomerases such as triosephosphateisomerase or bisphosphoglycerate mutase, intramolecular lyases such as adenylate cyclase and the hammerhead ribozyme, an RNA lyase. However, some enzymes that only have a single substrate do not fall into this category of mechanisms. Catalase is an example of this, as the enzyme reacts with a first molecule of hydrogen peroxide substrate, becomes oxidised and is then reduced by a second molecule of substrate. Although a single substrate is involved, the existence of a modified enzyme intermediate means that the mechanism of catalase is actually a ping–pong mechanism, a type of mechanism that is discussed in the *Multi-substrate reactions* section below.

Michaelis–Menten Kinetics

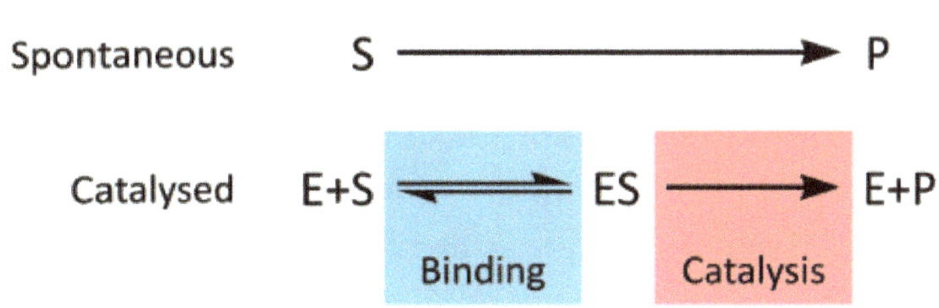

A chemical reaction mechanism with or without enzyme catalysis. The enzyme (E) binds substrate (S) to produce product (P).

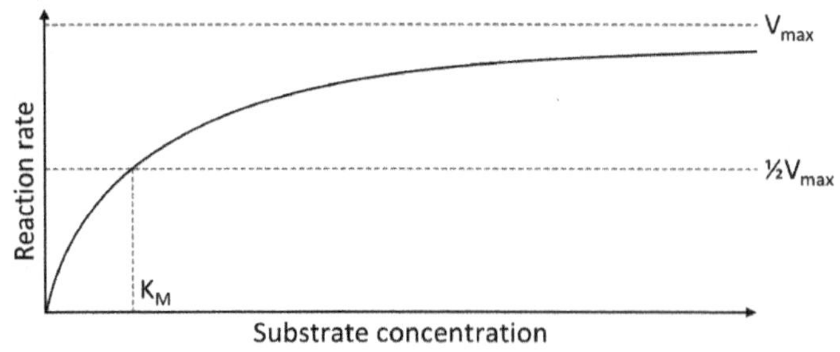

Saturation curve for an enzyme reaction showing the relation between the substrate concentration and reaction rate.

As enzyme-catalysed reactions are saturable, their rate of catalysis does not show a linear response to increasing substrate. If the initial rate of the reaction is measured over a range of substrate concentrations (denoted as [S]), the reaction rate (v) increases as [S] increases, as shown on the right. However, as [S] gets higher, the enzyme becomes saturated with substrate and the rate reaches V_{max}, the enzyme's maximum rate.

The Michaelis–Menten kinetic model of a single-substrate reaction is shown on the right. There is an initial bimolecular reaction between the enzyme E and substrate S to form the enzyme–substrate complex ES. The rate of enzymatic reaction increases with the increase of the substrate concentration up to a certain level called V_{max}; at V_{max}, increase in substrate concentration does not cause any increase in reaction rate as there no more enzyme (E) available for reacting with substrate (S). Here, the rate of reaction becomes dependent on the ES complex and the reaction becomes a unimolecular reaction with an order of zero. Though the enzymatic mechanism for the unimolecular reaction $\mathrm{ES} \xrightarrow{k_{cat}} \mathrm{E} + \mathrm{P}$ can be quite complex, there is typically one rate-determining enzymatic step that allows this reaction to be modelled as a single catalytic step with an apparent unimolecular rate constant k_{cat}. If the reaction path proceeds over one or several intermediates, k_{cat} will be a function of several elementary rate constants, whereas in the simplest case of a single elementary reaction (e.g. no intermediates) it will be identical to the elementary unimolecular rate constant k_2. The apparent unimolecular rate constant k_{cat} is also called *turnover number* and denotes the maximum number of enzymatic reactions catalysed per second.

The Michaelis–Menten equation describes how the (initial) reaction rate v_0 depends on the position of the substrate-binding equilibrium and the rate constant k_2.

$$v_0 = \frac{V_{\max}[\mathrm{S}]}{K_M + [\mathrm{S}]} \quad \textit{(Michaelis–Menten equation)}$$

with the constants

$$K_M \stackrel{\text{def}}{=} \frac{k_2 + k_{-1}}{k_1} \approx K_D$$

$$V_{\max} \stackrel{\text{def}}{=} k_{cat}[\mathrm{E}]_{tot}$$

This Michaelis–Menten equation is the basis for most single-substrate enzyme kinetics. Two crucial assumptions underlie this equation (apart from the general assumption about the mechanism only involving no intermediate or product inhibition, and there is no allostericity or cooperativity). The first assumption is the so-called quasi-steady-state assumption (or pseudo-steady-state hypothesis), namely that the concentration of the substrate-bound enzyme (and hence also the unbound enzyme) changes much more slowly than those of the product and substrate and thus the change over time of the complex can be set to zero $d[\mathrm{ES}]/dt \overset{!}{=} 0$. The second assumption is that the total enzyme concentration does not change over time, thus $[\mathrm{E}]_{\mathrm{tot}} = [\mathrm{E}] + [\mathrm{ES}] \overset{!}{=} \mathrm{const}$. A complete derivation can be found here. The Michaelis constant K_M is experimentally defined as the concentration at which the rate of the enzyme reaction is half V_{max}, which can be verified by substituting [S] = K_m into the Michaelis–Menten equation and can also be seen graphically. If the rate-determining enzymatic step is slow compared to substrate dissociation $(k_2 \ll k_{-1})$, the Michaelis constant K_M is roughly the dissociation constant K_D of the ES complex.

If [S] is small compared to K_M then the term $[\mathrm{S}]/(K_M + [\mathrm{S}]) \approx [\mathrm{S}]/K_M$ and also very little ES complex is formed, thus $[\mathrm{E}]_0 \approx [\mathrm{E}]$. Therefore, the rate of product formation is

$$v_0 \approx \frac{k_{cat}}{K_M}[\mathrm{E}][\mathrm{S}] \qquad \text{if } [\mathrm{S}] \ll K_M$$

Thus the product formation rate depends on the enzyme concentration as well as on the substrate concentration, the equation resembles a bimolecular reaction with a corresponding pseudo-second order rate constant k_2/K_M. This constant is a measure of catalytic efficiency. The most efficient enzymes reach a k_2/K_M in the range of $10^8 - 10^{10}\ \mathrm{M}^{-1}\ \mathrm{s}^{-1}$. These enzymes are so efficient they effectively catalyse a reaction each time they encounter a substrate molecule and have thus reached an upper theoretical limit for efficiency (diffusion limit); and are sometimes referred to as kinetically perfect enzymes.

Direct Use of the Michaelis–Menten Equation for Time Course Kinetic Analysis

The observed velocities predicted by the Michaelis–Menten equation can be used to directly model the time course disappearance of substrate and the production of product through incorporation of the Michaelis–Menten equation into the equation for first order chemical kinetics. This can only be achieved however if one recognises the problem associated with the use of Euler's number in the description of first order chemical kinetics. i.e. e^{-k} is a split constant that introduces a systematic error into calculations and can be rewritten as a single constant which represents the remaining substrate after each time period.

$$[S] = [S]_0(1-k)^t$$

$$[S] = [S]_0(1-v/[S]_0)^t$$

$$[S] = [S]_0(1-(V_{\max}[S]_0/(K_M+[S]_0)/[S]_0))^t$$

In 1983 Stuart Beal (and also independently Santiago Schnell and Claudio Mendoza in 1997) derived a closed form solution for the time course kinetics analysis of the Michaelis-Menten mechanism. The solution, known as the Schnell-Mendoza equation, has the form:

$$\frac{[S]}{K_M} = W[F(t)]$$

where W[] is the Lambert-W function.[14][15] and where F(t) is

$$F(t) = \frac{[S]_0}{K_M} \exp\left(\frac{[S]_0}{K_M} - \frac{V_{\max}}{K_M}t\right)$$

This equation is encompassed by the equation below, obtained by Berberan-Santos (MATCH Commun. Math. Comput. Chem. 63 (2010) 283), which is also valid when the initial substrate concentration is close to that of enzyme,

$$\frac{[S]}{K_M} = W[F(t)] - \frac{V_{\max}}{k_{cat}K_M}\frac{W[F(t)]}{1+W[F(t)]}$$

where W[] is again the Lambert-W function.

Linear Plots of the Michaelis–Menten Equation

The plot of v versus [S] above is not linear; although initially linear at low [S], it bends over to saturate at high [S]. Before the modern era of nonlinear curve-fitting on computers, this nonlinearity could make it difficult to estimate K_M and V_{max} accurately. Therefore, several researchers developed linearisations of the Michaelis–Menten equation, such as the Lineweaver–Burk plot, the Eadie–Hofstee diagram and the Hanes–Woolf plot. All of these linear representations can be useful for visualising data, but none should be used to determine kinetic parameters, as computer software is readily available that allows for more accurate determination by nonlinear regression methods.

The Lineweaver–Burk plot or double reciprocal plot is a common way of illustrating kinetic data. This is produced by taking the reciprocal of both sides of the Michaelis–Menten equation. As shown on the right, this is a linear form of the Michaelis–Menten equation and produces a straight line with the equation $y = mx + c$ with a y-intercept equivalent to $1/V_{max}$ and an x-intercept of the graph representing $-1/K_M$.

$$\frac{1}{v} = \frac{K_M}{V_{\max}[S]} + \frac{1}{V_{\max}}$$

Naturally, no experimental values can be taken at negative 1/[S]; the lower limiting value 1/[S] = 0 (the y-intercept) corresponds to an infinite substrate concentration, where $1/v=1/V_{max}$ as shown at the right; thus, the x-intercept is an extrapolation of the experimental data taken at positive concentrations. More generally, the Lineweaver–Burk plot skews the importance of measurements taken at low substrate concentrations and, thus, can yield inaccurate estimates of V_{max} and K_M. A more accurate linear plotting method is the Eadie-Hofstee plot. In this case, v is plotted against v/[S]. In the third common linear representation, the Hanes-Woolf plot, [S]/v is plotted against [S]. In general, data normalisation can help diminish the amount of experimental work and can increase the reliability of the output, and is suitable for both graphical and numerical analysis.

Practical Significance of Kinetic Constants

The study of enzyme kinetics is important for two basic reasons. Firstly, it helps explain how enzymes work, and secondly, it helps predict how enzymes behave in living organisms. The kinetic constants defined above, K_M and V_{max}, are critical to attempts to understand how enzymes work together to control metabolism.

Making these predictions is not trivial, even for simple systems. For example, oxaloacetate is formed by malate dehydrogenase within the mitochondrion. Oxaloacetate can then be consumed by citrate synthase, phosphoenolpyruvate carboxykinase or aspartate aminotransferase, feeding into the citric acid cycle, gluconeogenesis or aspartic acid biosynthesis, respectively. Being able to predict how much oxaloacetate goes into which pathway requires knowledge of the concentration of oxaloacetate as well as the concentration and kinetics of each of these enzymes. This aim of predicting the behaviour of metabolic pathways reaches its most complex expression in the synthesis of huge amounts of kinetic and gene expression data into mathematical models of entire organisms. Alternatively, one useful simplification of the metabolic modelling problem is to ignore the underlying enzyme kinetics and only rely on information about the reaction network's stoichiometry, a technique called flux balance analysis.

Michaelis–Menten Kinetics with Intermediate

One could also consider the less simple case

$$\mathrm{E} + \mathrm{S} \underset{k_{-1}}{\overset{k_1}{\rightleftharpoons}} \mathrm{ES} \overset{k_2}{\longrightarrow} \mathrm{EI} \overset{k_3}{\longrightarrow} \mathrm{E} + \mathrm{P}$$

where a complex with the enzyme and an intermediate exists and the intermediate is converted into product in a second step. In this case we have a very similar equation

$$v_0 = k_{cat} \frac{[\mathrm{S}][\mathrm{E}]_0}{K'_M + [\mathrm{S}]}$$

but the constants are different

$$K'_M \overset{\text{def}}{=} \frac{k_3}{k_2 + k_3} K_M = \frac{k_3}{k_2 + k_3} \cdot \frac{k_2 + k_{-1}}{k_1}$$

$$k_{cat} \overset{\text{def}}{=} \frac{k_3 k_2}{k_2 + k_3}$$

We see that for the limiting case $k_3 \gg k_2$, thus when the last step from $\mathrm{EI} \longrightarrow \mathrm{E} + \mathrm{P}$ is much faster than the previous step, we get again the original equation. Mathematically we have then $K'_M \approx K_M$ and $k_{cat} \approx k_2$.

Multi-Substrate Reactions

Multi-substrate reactions follow complex rate equations that describe how the substrates bind and in what sequence. The analysis of these reactions is much simpler if the concentration of substrate A is kept constant and substrate B varied. Under these conditions, the enzyme behaves just like a single-substrate enzyme and a plot of v by [S] gives apparent K_M and V_{max} constants for substrate B. If a set of these measurements is performed at different fixed concentrations of A, these data can be used to work out what the mechanism of the reaction is. For an enzyme that takes two substrates A and B and turns them into two products P and Q, there are two types of mechanism: ternary complex and ping–pong.

Ternary-Complex Mechanisms

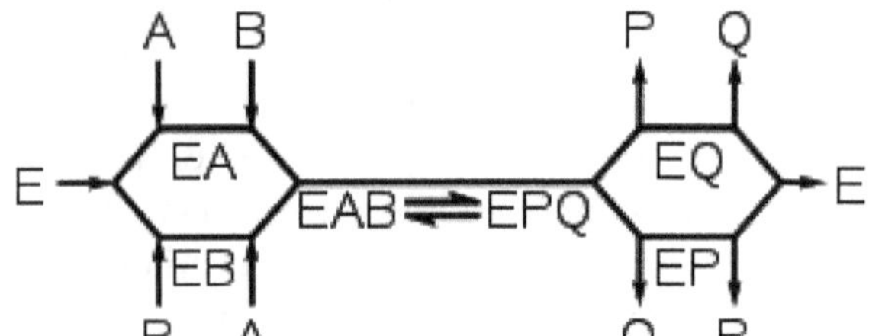

Random-order ternary-complex mechanism for an enzyme reaction. The reaction path is shown as a line and enzyme intermediates containing substrates A and B or products P and Q are written below the line.

In these enzymes, both substrates bind to the enzyme at the same time to produce an EAB ternary complex. The order of binding can either be random (in a random mechanism) or substrates have to bind in a particular sequence (in an ordered mechanism). When a set of v by [S] curves (fixed A, varying B) from an enzyme with a ternary-complex mechanism are plotted in a Lineweaver–Burk plot, the set of lines produced will intersect.

Enzymes with ternary-complex mechanisms include glutathione *S*-transferase, dihydrofolate reductase and DNA polymerase.The following links show short animations of the ternary-complex mechanisms of the enzymes dihydrofolate reductase[β] and DNA polymerase[γ].

Ping–Pong Mechanisms

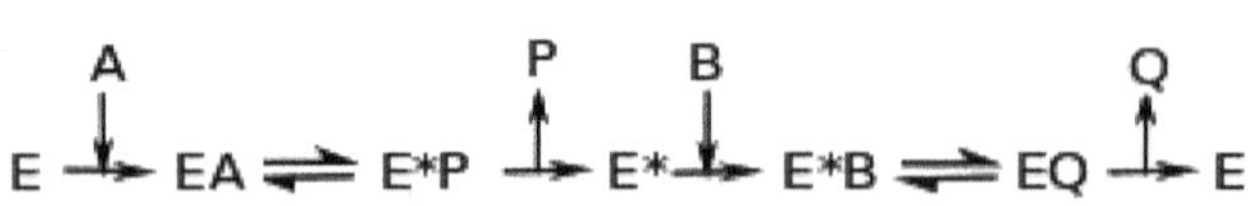

Ping–pong mechanism for an enzyme reaction. Intermediates contain substrates A and B or products P and Q.

As shown on the right, enzymes with a ping-pong mechanism can exist in two states, E and a chemically modified form of the enzyme E*; this modified enzyme is known as an intermediate. In such mechanisms, substrate A binds, changes the enzyme to E* by, for example, transferring a chemical group to the active site, and is then released. Only after the first substrate is released can substrate B bind and react with the modified enzyme, regenerating the unmodified E form. When a set of v by [S] curves (fixed A, varying B) from an enzyme with a ping–pong mechanism are plotted in a Lineweaver–Burk plot, a set of parallel lines will be produced. This is called a secondary plot. Enzymes with ping–pong mechanisms include some oxidoreductases such as thioredoxin peroxidase, transferases such as acylneuraminate cytidylyltransferase and serine proteases such as trypsin and chymotrypsin. Serine proteases are a very common and diverse family of enzymes, including digestive enzymes (trypsin, chymotrypsin, and elastase), several enzymes of the blood clotting cascade and many others. In these serine proteases, the E* intermediate is an acyl-enzyme species formed by the attack of an active site serine residue on a peptide bond in a protein substrate. A short animation showing the mechanism of chymotrypsin is linked here.[δ]

NON-MICHAELIS–MENTEN KINETICS

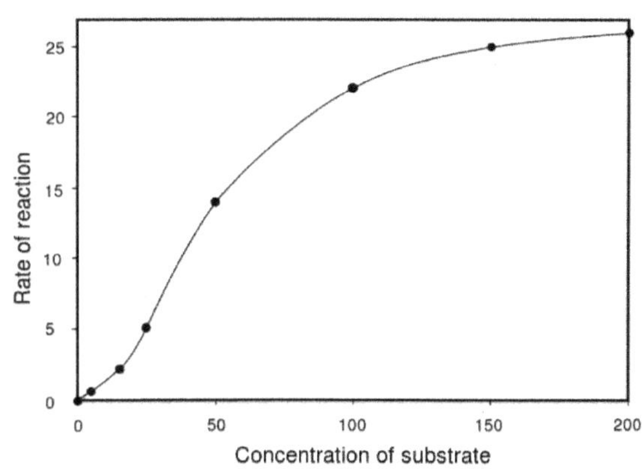

Saturation curve for an enzyme reaction showing sigmoid kinetics.

Some enzymes produce a sigmoid *v* by [S] plot, which often indicates cooperative binding of substrate to the active site. This means that the binding of one substrate molecule affects the binding of subsequent substrate molecules. This behavior is most common in multimeric enzymes with several interacting active sites. Here, the mechanism of cooperation is similar to that of hemoglobin, with binding of substrate to one active site altering the affinity of the other active sites for substrate molecules. Positive cooperativity occurs when binding of the first substrate molecule *increases* the affinity of the other active sites for substrate. Negative cooperativity occurs when binding of the first substrate *decreases* the affinity of the enzyme for other substrate molecules.

Allosteric enzymes include mammalian tyrosyl tRNA-synthetase, which shows negative cooperativity, and bacterial aspartate transcarbamoylase and phosphofructokinase, which show positive cooperativity.

Cooperativity is surprisingly common and can help regulate the responses of enzymes to changes in the concentrations of their substrates. Positive cooperativity makes enzymes much more sensitive to [S] and their activities can show large changes over a narrow range of substrate concentration. Conversely, negative cooperativity makes enzymes insensitive to small changes in [S].

The Hill equation (biochemistry) is often used to describe the degree of cooperativity quantitatively in non-Michaelis–Menten kinetics. The derived Hill coefficient n measures how much the binding of substrate to one active site affects the binding of substrate to the other active sites. A Hill coefficient of <1 indicates negative cooperativity and a coefficient of >1 indicates positive cooperativity.

Pre-steady-State Kinetics

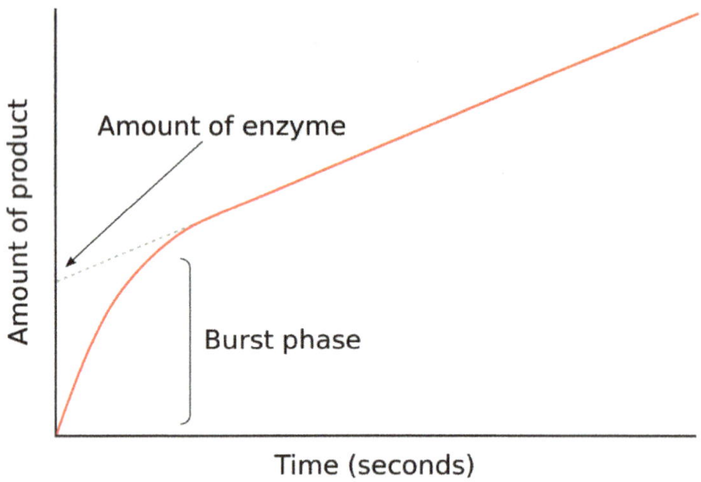

Pre-steady state progress curve, showing the burst phase of an enzyme reaction.

In the first moment after an enzyme is mixed with substrate, no product has been formed and no intermediates exist. The study of the next few milliseconds of the reaction is called pre-steady-state kinetics. Pre-steady-state kinetics is therefore concerned with the formation and consumption of enzyme–substrate intermediates (such as ES or E*) until their steady-state concentrations are reached.

This approach was first applied to the hydrolysis reaction catalysed by chymotrypsin. Often, the detection of an intermediate is a vital piece of evidence in investigations of what mechanism an enzyme follows. For example, in the ping–pong mechanisms that are shown above, rapid kinetic measurements can follow the release of product P and measure the formation of the modified enzyme intermediate E*. In the case of chymotrypsin, this intermediate is formed by an attack on the substrate by the nucleophilic serine in the active site and the formation of the acyl-enzyme intermediate.

In the figure to the right, the enzyme produces E* rapidly in the first few seconds of the reaction. The rate then slows as steady state is reached. This rapid burst phase of the reaction measures a single turnover of the enzyme. Consequently, the amount of product released in this burst, shown as the intercept on the y-axis of the graph, also gives the amount of functional enzyme which is present in the assay.

Chemical Mechanism

An important goal of measuring enzyme kinetics is to determine the chemical mechanism of an enzyme reaction, i.e., the sequence of chemical steps that transform substrate into product. The kinetic approaches discussed above will show at what rates intermediates are formed and inter-converted, but they cannot identify exactly what these intermediates are.

Kinetic measurements taken under various solution conditions or on slightly modified enzymes or substrates often shed light on this chemical mechanism, as they reveal the rate-determining step or intermediates in the reaction. For example, the breaking of a covalent bond to a hydrogen atom is a common rate-determining step. Which of the possible hydrogen transfers is rate determining can be shown by measuring the kinetic effects of substituting each hydrogen by deuterium, its stable isotope. The rate will change when the critical hydrogen is replaced, due to a primary kinetic isotope effect, which occurs because bonds to deuterium are harder to break than bonds to hydrogen. It is also possible to measure similar effects with other isotope substitutions, such as $^{13}C/^{12}C$ and $^{18}O/^{16}O$, but these effects are more subtle.

Isotopes can also be used to reveal the fate of various parts of the substrate molecules in the final products. For example, it is sometimes difficult to discern the origin of an oxygen atom in the final product; since it may have come from water or from part of the substrate. This may be determined by systematically substituting oxygen's stable isotope ^{18}O into the various molecules that participate in the reaction and checking for the isotope in the product. The chemical mechanism can also be elucidated by examining the kinetics and isotope effects under different pH conditions, by altering the metal ions or other bound cofactors, by site-directed mutagenesis of conserved amino acid residues, or by studying the behaviour of the enzyme in the presence of analogues of the substrate(s).

ENZYME INHIBITION AND ACTIVATION

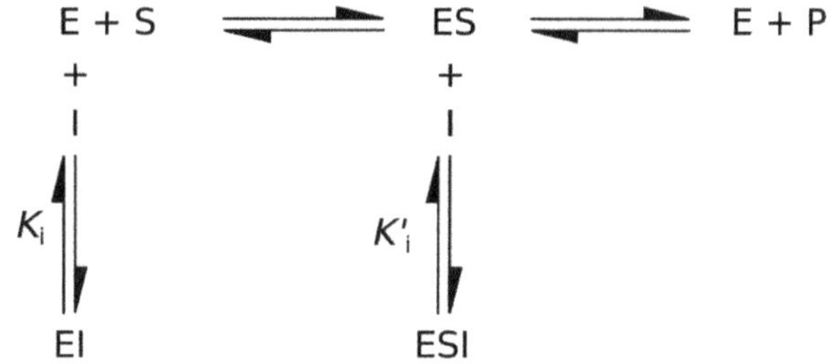

Kinetic scheme for reversible enzyme inhibitors.

Enzyme inhibitors are molecules that reduce or abolish enzyme activity, while enzyme activators are molecules that increase the catalytic rate of enzymes. These interactions can be either *reversible* (i.e., removal of the inhibitor restores enzyme activity) or *irreversible* (i.e., the inhibitor permanently inactivates the enzyme).

Reversible Inhibitors

Traditionally reversible enzyme inhibitors have been classified as competitive, uncompetitive, or non-competitive, according to their effects on K_m and V_{max}. These different effects result from the inhibitor binding to the enzyme E, to the enzyme–substrate complex ES, or to both, respectively. The division of these classes arises from a problem in their derivation and results in the need to use two different binding constants for one binding event. The binding of an inhibitor and its effect on the enzymatic activity are two distinctly different things, another problem the traditional equations fail to acknowledge. In noncompetitive inhibition the binding of the inhibitor results in 100% inhibition of the enzyme only, and fails to consider the possibility of anything in between. The common form of the inhibitory term also obscures the relationship between the inhibitor binding to the enzyme and its relationship to any other binding term be it the Michaelis–Menten equation or a dose response curve associated with ligand receptor binding. To demonstrate the relationship the following rearrangement can be made:

$$\frac{V_{max}}{1+\frac{[I]}{K_i}} = \frac{V_{max}}{\frac{[I]+K_i}{K_i}}$$

Adding zero to the bottom ([I]-[I])

$$\frac{V_{max}}{\frac{[I]+K_i}{[I]+K_i-[I]}}$$

Dividing by [I]+K$_i$

$$\frac{V_{max}}{\frac{1}{1-\frac{[I]}{[I]+K_i}}} = V_{max} - V_{max}\frac{[I]}{[I]+K_i}$$

This notation demonstrates that similar to the Michaelis–Menten equation, where the rate of reaction depends on the percent of the enzyme population interacting with substrate

fraction of the enzyme population bound by substrate

$$\frac{[S]}{[S]+K_m}$$

fraction of the enzyme population bound by inhibitor

$$\frac{[I]}{[I]+K_i}$$

the effect of the inhibitor is a result of the percent of the enzyme population interacting with inhibitor. The only problem with this equation in its present form is that it assumes absolute inhibition of the enzyme with inhibitor binding, when in fact there can be a wide range of effects anywhere from 100% inhibition of substrate turn over to just >0%. To account for this the equation can be easily modified to allow for different degrees of inhibition by including a delta V_{max} term.

$$V_{\max} - \Delta V_{\max}\frac{[I]}{[I]+K_i}$$

or

$$V_{\max 1} - (V_{\max 1} - V_{\max 2})\frac{[I]}{[I]+K_i}$$

This term can then define the residual enzymatic activity present when the inhibitor is interacting with individual enzymes in the population. However the inclusion of this term has the added value of allowing for the possibility of activation if the secondary V_{max} term turns out to be higher than the initial term. To account for the possibly of activation as well the notation can then be rewritten replacing the inhibitor "I" with a modifier term denoted here as "X".

$$V_{\max 1} - (V_{\max 1} - V_{\max 2})\frac{[X]}{[X]+K_x}$$

While this terminology results in a simplified way of dealing with kinetic effects relating to the maximum velocity of the Michaelis–Menten equation, it highlights potential problems with the term used to describe effects relating to the K_m. The K_m relating to the affinity of the enzyme for the substrate should in most cases relate to potential changes in the binding site of the enzyme which would directly result from enzyme inhibitor interactions. As such a term similar to the one proposed above to modulate V_{max} should be appropriate in most situations:

$$K_{m1} - (K_{m1} - K_{m2})\frac{[X]}{[X]+K_x}$$

Irreversible Inhibitors

Enzyme inhibitors can also irreversibly inactivate enzymes, usually by covalently modifying active site residues. These reactions, which may be called suicide substrates, follow exponential decay functions and are usually saturable. Below saturation, they follow first order kinetics with respect to inhibitor.

Mechanisms of Catalysis

The favoured model for the enzyme–substrate interaction is the induced fit model. This model proposes that the initial interaction between enzyme and substrate is relatively weak, but that these weak interactions rapidly induce conformational changes in the enzyme that strengthen binding. These conformational changes also bring catalytic residues in the active site close to the chemical bonds in the substrate that will be altered in the reaction. Conformational changes can be measured using circular dichroism or dual polarisation interferometry. After binding takes place, one or more mechanisms of catalysis lower the energy of the reaction's transition state by providing an alternative chemical pathway for the reaction. Mechanisms of catalysis include catalysis by bond strain; by proximity and orientation; by active-site proton donors or acceptors; covalent catalysis and quantum tunnelling.

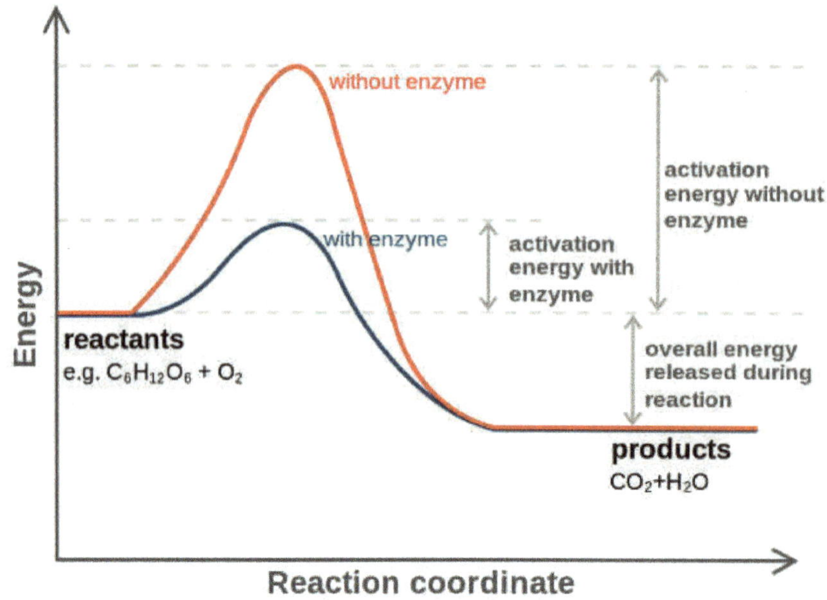

The energy variation as a function of reaction coordinate shows the stabilisation of the transition state by an enzyme.

Enzyme kinetics cannot prove which modes of catalysis are used by an enzyme. However, some kinetic data can suggest possibilities to be examined by other techniques. For example, a ping–pong mechanism with burst-phase pre-steady-state kinetics would suggest covalent catalysis might be important in this enzyme's mechanism. Alternatively, the observation of a strong pH effect on V_{max} but not K_m might indicate that a residue in the active site needs to be in a particular ionisation state for catalysis to occur.

In 1902 Victor Henri proposed a quantitative theory of enzyme kinetics, but at the time the experimental significance of the hydrogen ion concentration was not yet recognized. After Peter Lauritz Sørensen had defined the logarithmic pH-scale and introduced the concept of buffering in 1909 the German chemist Leonor Michaelis and his Canadian postdoc Maud Leonora Menten repeated Henri's experiments and confirmed his equation, which is now generally referred to as Michaelis-Menten kinetics (sometimes also *Henri-Michaelis-Menten kinetics*). Their work was further developed by G. E. Briggs and J. B. S. Haldane, who derived kinetic equations that are still widely considered today a starting point in modeling enzymatic activity.

The major contribution of the Henri-Michaelis-Menten approach was to think of enzyme reactions in two stages. In the first, the substrate binds reversibly to the enzyme, forming the enzyme-substrate complex. This is sometimes called the Michaelis complex. The enzyme then catalyzes the chemical step in the reaction and releases the product. The kinetics of many enzymes is adequately described by the simple Michaelis-Menten model, but all enzymes have internal motions that are not accounted for in the model and can have significant contributions to the overall reaction kinetics. This can be modeled by introducing several Michaelis-Menten pathways that are connected with fluctuating rates, which is a mathematical extension of the basic Michaelis Menten mechanism.

ALLOSTERIC REGULATION

In biochemistry, **allosteric regulation** (or **allosteric control**) is the regulation of an enzyme by binding an effector molecule at a site other than the enzyme's active site.

The site to which the effector binds is termed the *allosteric site*. Allosteric sites allow effectors to bind to the protein, often resulting in a conformational change involving protein dynamics. Effectors that enhance the protein's activity are referred to as *allosteric activators*, whereas those that decrease the protein's activity are called *allosteric inhibitors*.

Allosteric regulations are a natural example of control loops, such as feedback from downstream products or feedforward from upstream substrates. Long-range allostery is especially important in cell signaling. Allosteric regulation is also particularly important in the cell's ability to adjust enzyme activity.

The term *allostery* comes from the Greek *allos*, solid (object). This is in reference to the fact that the regulatory site of an allosteric protein is physically distinct from its active site.

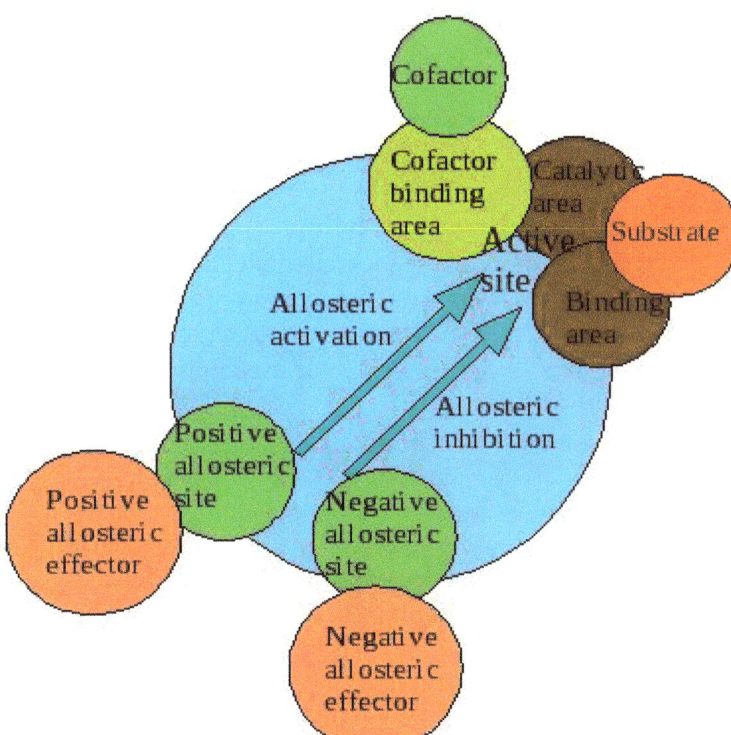

Allosteric regulation of an enzyme

Models of Allosteric Regulation

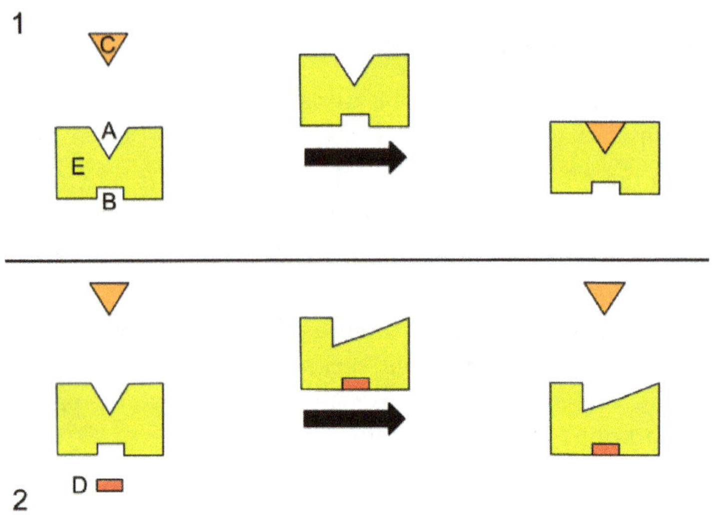

A - Active site

B - Allosteric site

C - Substrate

D - Inhibitor

E - Enzyme

This is a diagram of allosteric regulation of an enzyme.

Most allosteric effects can be explained by the *concerted* MWC model put forth by Monod, Wyman, and Changeux, or by the sequential model described by Koshland, Nemethy, and Filmer. Both postulate that enzyme subunits exist in one of two conformations, tensed (T) or relaxed (R), and that relaxed subunits bind substrate more readily than those in the tense state. The two models differ most in their assumptions about subunit interaction and the preexistence of both states.

Concerted Model

The concerted model of allostery, also referred to as the symmetry model or MWC model, postulates that enzyme subunits are connected in such a way that a conformational change in one subunit is necessarily conferred to all other subunits. Thus, all subunits must exist in the same conformation. The model further holds that, in the absence of any ligand (substrate or otherwise), the equilibrium favors one of the conformational states, T or R. The equilibrium can be shifted to the R or T state through the binding of one ligand (the allosteric effector or ligand) to a site that is different from the active site (the allosteric site).

Sequential Model

The sequential model of allosteric regulation holds that subunits are not connected in such a way that a conformational change in one induces a similar change in the others. Thus, all enzyme subunits do not necessitate the same conformation. Moreover, the sequential model dictates that molecules of a substrate bind via an induced fit protocol. In general, when a subunit randomly collides with a molecule of substrate, the active site, in essence, forms a glove around its substrate. While such an induced fit converts a subunit from the tensed state to relaxed state, it does not propagate the conformational change to adjacent subunits. Instead, substrate-binding at one subunit only slightly alters the structure of other subunits so that their binding sites are more receptive to substrate. To summarize:

- subunits need not exist in the same conformation
- molecules of substrate bind via induced-fit protocol
- conformational changes are not propagated to all subunits

Morpheein Model

The morpheein model of allosteric regulation is a dissociative concerted model.

A morpheein is a homo-oligomeric structure that can exist as an ensemble of physiologically significant and functionally different alternate quaternary assemblies. Transitions between alternate morpheein assemblies involve oligomer dissociation, conformational change in the dissociated state, and reassembly to a different oligomer. The required oligomer disassembly step differentiates the morpheein model for allosteric regulation from the classic MWC and KNF models. Porphobilinogen synthase (PBGS) is the prototype morpheein.

Ensemble Models

Ensemble models of allosteric regulation enumerate an allosteric system's statistical ensemble as a function of its energy functional, and then relate specific statistical measurements of allostery to specific energy terms in the the energy function (such as an intermolecular salt bridge between two domains). Ensemble models like the Ensemble Allosteric Model and Allosteric Ising Model assume that each domain of the system can adopt two states similar to the MWC model. The allostery landscape model introduced by Cuendet, Weinstein, and LeVine allows for the domains to have any number of states and the contribution of a specific molecular interaction to a given allosteric coupling can be estimated using a rigorous set of rules. Molecular dynamics simulations can be used to estimate a systems's statistical ensemble so that it can be analyzed with the allostery landscape model.

Allosteric Resources Online

Allosteric Database

Allostery is a direct and efficient means for regulation of biological macromolecule function, produced by the binding of a ligand at an allosteric site topographically distinct from the orthosteric site. Due to the often high receptor selectivity and lower target-based toxicity, allosteric regulation is also expected to play an increasing role in drug discovery and bioengineering. The AlloSteric Database (ASD, http://mdl.shsmu.edu.cn/ASD) provides a central resource for the display, search and analysis of the structure, function and related annotation for allosteric molecules. Currently, ASD contains allosteric proteins from more than 100 species and modulators in three categories (activators, inhibitors, and regulators). Each protein is annotated with detailed description of allostery, biological process and related diseases, and each modulator with binding affinity, physicochemical properties and therapeutic area. Integrating the information of allosteric proteins in ASD should allow the prediction of allostery for unknown proteins, to be followed with experimental validation. In addition, modulators curated in ASD can be used to investigate potential allosteric targets for a query compound, and can help chemists to implement structure modifications for novel allosteric drug design.

Allosteric Residues & their Prediction using the STRESS Web Server

Not all protein residues play equally important roles in allosteric regulation. The identification of residues that are essential to allostery (so-called "allosteric residues") has been the focus of many studies, especially within the last decade. In part, this growing interest is a result of their general importance in protein science, but also because allosteric residues may be exploited in biomedical contexts. Pharmacologically important proteins with difficult-to-target sites may yield to approaches in which one alternatively targets easier-to-reach residues that are capable of allosterically regulating the primary site of interest. These residues can broadly be classified as surface- and interior-allosteric amino acids. Allosteric sites at the surface generally play regulatory roles that are fundamentally distinct from those within the interior; surface residues may serve as receptors or effector sites in allosteric signal transmission, whereas those within the interior may act to transmit such signals. STRucturally-identified ESSential residues (STRESS, http://stress.molmovdb.org) is a web tool that enables users to submit their own protein structures of interest in order to predict both surface- and interior-allosteric residues in an algorithmically efficient manner. The software behind this server employs 3D structures to build models of conformational change in order to perform predictions.

Allosteric Modulation

Positive Modulation

Positive allosteric modulation (also known as *allosteric activation*) occurs when the binding of one ligand enhances the attraction between substrate molecules and other binding sites. An example is the binding of oxygen molecules to hemoglobin, where oxygen is effectively both the substrate and the effector. The allosteric, or "other", site is the active site of an adjoining protein subunit. The binding of oxygen to one subunit induces a conformational change in that subunit that interacts with the remaining active sites to enhance *their* oxygen affinity. Another example of allosteric activation is seen in cytosolic IMP-GMP specific 5'-nucleotidase II (cN-II), where the affinity for substrate GMP increases upon GTP binding at the dimer interface.

Negative Modulation

Negative allosteric modulation (also known as *allosteric inhibition*) occurs when the binding of one ligand decreases the affinity for substrate at other active sites. For example, when 2,3-BPG binds to an allosteric site on hemoglobin, the affinity for oxygen of all subunits decreases. This is when a regulator is absent from the binding site.

Direct thrombin inhibitors provides an excellent example of negative allosteric modulation. Allosteric inhibitors of thrombin have been discovered which could potentially be used as anticoagulants.

Another example is strychnine, a convulsant poison, which acts as an allosteric inhibitor of the glycine receptor. Glycine is a major post-synaptic inhibitory neurotransmitter in mammalian spinal cord and brain stem. Strychnine acts at a separate binding site on the glycine receptor in an allosteric manner; i.e., its binding lowers the affinity of the glycine receptor for glycine. Thus, strychnine inhibits the action of an inhibitory transmitter, leading to convulsions.

Another instance in which negative allosteric modulation can be seen is between ATP and the enzyme Phosphofructokinase within the negative feedback loop that regulates glycolysis. Phosphofructokinase (generally referred to as PFK) is an enzyme that catalyses the third step of glycolysis: the phosphorylation of Fructose-6-phosphate into Fructose 1,6-bisphosphate. PFK can be allosterically inhibited by high levels of ATP within the cell. When ATP levels are high, ATP will bind to an allosteoric site on phosphofructokinase, causing a change in the enzyme's three-dimensional shape. This change causes its affinity for substrate (fructose-6-phosphate and ATP) at the active site to decrease, and the enzyme is deemed inactive. This causes glycolysis to cease when ATP levels are high, thus conserving the body's glucose and maintaining balanced levels of cellular ATP. In this way, ATP serves as a negative allosteric modulator for PFK, despite the fact that it is also a substrate of the enzyme.

Types of Allosteric Regulation

Homotropic

A homotropic allosteric modulator is a substrate for its target enzyme, as well as a regulatory molecule of the enzyme's activity. It is typically an activator of the enzyme. For example, O_2 and CO are homotropic allosteric modulators of hemoglobin.

Heterotropic

A heterotropic allosteric modulator is a regulatory molecule that is not the enzyme's substrate. It may be either an activator or an inhibitor of the enzyme. For example, H^+,

CO_2, and 2,3-bisphosphoglycerate are heterotropic allosteric modulators of hemoglobin. Some allosteric proteins can be regulated by both their substrates and other molecules. Such proteins are capable of both homotropic and heterotropic interactions.

Non-Regulatory Allostery

A non-regulatory allosteric site refers to any non-regulatory component of an enzyme (or any protein), that is not itself an amino acid. For instance, many enzymes require sodium binding to ensure proper function. However, the sodium does not necessarily act as a regulatory subunit; the sodium is always present and there are no known biological processes to add/remove sodium to regulate enzyme activity. Non-regulatory allostery could comprise any other ions besides sodium (calcium, magnesium, zinc), as well as other chemicals and possibly vitamins.

Pharmacology

Allosteric modulation of a receptor results from the binding of allosteric modulators at a different site (a "regulatory site") from that of the endogenous ligand (an "active site") and enhances or inhibits the effects of the endogenous ligand. Under normal circumstances, it acts by causing a conformational change in a receptor molecule, which results in a change in the binding affinity of the ligand. In this way, an allosteric ligand modulates the receptor's activation by its primary (orthosteric) ligand, and can be thought to act like a dimmer switch in an electrical circuit, adjusting the intensity of the response.

For example, the $GABA_A$ receptor has two active sites that the neurotransmitter gamma-aminobutyric acid (GABA) binds, but also has benzodiazepine and general anaesthetic agent regulatory binding sites. These regulatory sites can each produce positive allosteric modulation, potentiating the activity of GABA. Diazepam is an agonist at the benzodiazepine regulatory site, and its antidote flumazenil is an antagonist.

More recent examples of drugs that allosterically modulate their targets include the calcium-mimicking cinacalcet and the HIV treatment maraviroc.

Allosteric Sites as Drug Targets

Allosteric sites may represent a novel drug target. There are a number of advantages in using allosteric modulators as preferred therapeutic agents over classic orthosteric ligands. For example, G protein-coupled receptor (GPCR) allosteric binding sites have not faced the same evolutionary pressure as orthosteric sites to accommodate an endogenous ligand, so are more diverse. Therefore, greater GPCR selectivity may be obtained by targeting allosteric sites. This is particularly useful for GPCRs where selective orthosteric therapy has been difficult because of sequence conservation of the orthosteric site across receptor subtypes. Also, these modulators have a decreased potential for toxic effects, since modulators with limited co-operativity will have a ceiling level to their effect, irrespective of the administered dose. Another type of pharmacological selectivity that is unique to allosteric modulators is based on co-operativity. An allosteric modulator may display neutral co-operativity with an orthosteric ligand at all subtypes of a given receptor except the subtype of interest, which is termed "absolute subtype selectivity". If an allosteric modulator does not possess appreciable efficacy, it can provide another powerful therapeutic advantage over orthosteric ligands, namely the ability to selectively tune up or down tissue responses only when the endogenous agonist is present. Oligomer-specific small molecule binding sites are drug targets for medically relevant morpheeins.

SOFTWARE

ENZO

ENZO (Enzyme Kinetics) is a graphical interface tool for building kinetic models of enzyme catalyzed reactions. ENZO automatically generates the corresponding differential equations from a stipulated enzyme reaction scheme. These differential equations are processed by a numerical solver and a regression algorithm which fits the coefficients of differential equations to experimentally observed time course curves. ENZO allows rapid evaluation of rival reaction schemes and can be used for routine tests in enzyme kinetics.

PROTEIN DYNAMICS

Proteins are generally thought to adopt unique structures determined by their amino acid sequences, as outlined by Anfinsen's dogma. However, proteins are not strictly static objects, but rather populate ensembles of (sometimes similar) conformations. Transitions between these states occur on a variety of length scales (tenths of Å to nm) and time scales (ns to s), and have been linked to functionally relevant phenomena such as allosteric signaling and enzyme catalysis.

The study of protein dynamics is most directly concerned with the transitions between these states, but can also involve the nature and equilibrium populations of the states themselves. These two perspectives—kinetics and thermodynamics, respectively—can be conceptually synthesized in an "energy landscape" paradigm: highly populated states and the kinetics of transitions between them can be described by the depths of energy wells and the heights of energy barriers, respectively.

Local Flexibility: Atoms and Residues

Portions of protein structures often deviate from the equilibrium state. Some such excursions are harmonic, such as stochastic fluctuations of chemical bonds and bond angles. Others are anharmonic, such as sidechains that jump between separate discrete energy minima, or rotamers. Evidence for local flexibility is often obtained from NMR spectroscopy. Flexible and potentially disordered regions of a protsein can be detected using the random coil index. Flexibility in folded proteins can be identified by analyzing the spin relaxation of individual atoms in the protein. Flexibility can also be observed in very high-resolution electron density maps produced by X-ray crystallography, particularly when diffraction data is collected at room temperature instead of the traditional cryogenic temperature (typically near 100 K).

Regional Flexibility: Intra-Domain Multi-Residue Coupling

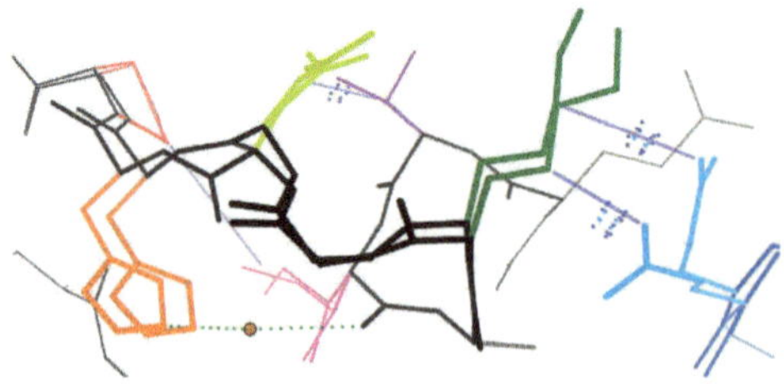

A network of alternative conformations in catalase (Protein Data Bank code: 1gwe) with diverse properties. Multiple phenomena define the network: van der Waals interactions (blue dots and line segments) between sidechains, a hydrogen bond (dotted green line) through a partial-occupancy water (brown), coupling through the locally mobile backbone (black), and perhaps electrostatic forces between the Lys (green) and nearby polar residues (blue: Glu, yellow: Asp, purple: Ser). This particular network is distal from the active site and is therefore putatively not critical for function.

Many residues are in close spatial proximity in protein structures. This is true for most residues that are contiguous in the primary sequence, but also for many that are distal in sequence yet are brought into contact in the final folded structure. Because of this proximity, these residues's energy landscapes become coupled based on various biophysical phenomena such as hydrogen bonds, ionic bonds, and van der Waals interactions (see figure). Transitions between states for such sets of residues therefore become correlated.

This is perhaps most obvious for surface-exposed loops, which often shift collectively to adopt different conformations in different crystal structures (see figure). However, coupled conformational heterogeneity is also sometimes evident in secondary structure. For example, consecutive residues and residues offset by 4 in the primary sequence often interact in α helices. Also, residues offset by 2 in the primary sequence point their sidechains toward the same face of β sheets and are close enough to interact sterically, as are residues on adjacent strands of the same β sheet.

An "ensemble" of 44 crystal structures of hen egg white lysozyme from the Protein Data Bank, showing that different crystallization conditions lead to different conformations for various surface-exposed loops and termini (red arrows).

When these coupled residues form pathways linking functionally important parts of a protein, they may participate in allosteric signaling. For example, when a molecule of oxygen binds to one subunit of the hemoglobin tetramer, that information is allosterically propagated to the other three subunits, thereby enhancing their affinity for oxygen. In this case, the coupled flexibility in hemoglobin allows for cooperative oxygen binding, which is physiologically useful because it allows rapid oxygen loading in lung tissue and rapid oxygen unloading in oxygen-deprived tissues (e.g. muscle).

Global Flexibility: Multiple Domains

The presence of multiple domains in proteins gives rise to a great deal of flexibility and mobility, leading to **protein domain dynamics**. Domain motions can be inferred by comparing different structures of a protein (as in Database of Molecular Motions), or they can be directly observed using spectra measured by neutron spin echo spectroscopy. They can also be suggested by sampling in extensive molecular dynamics trajectories and principal component analysis. Domain motions are important for:

- catalysis
- regulatory activity
- transport of metabolites
- formation of protein assemblies
- cellular locomotion

One of the largest observed domain motions is the 'swivelling' mechanism in pyruvate phosphate dikinase. The phosphoinositide domain swivels between two states in order to bring a phosphate group from the active site of the nucleotide binding domain to that of the phosphoenolpyruvate/pyruvate domain. The phosphate group is moved over a distance of 45 Å involving a domain motion of about 100 degrees around a single residue. In enzymes, the closure of one domain onto another captures a substrate by an induced fit, allowing the reaction to take place in a controlled way. A detailed analysis by Gerstein led to the classification of two basic types of domain motion; hinge and shear. Only a relatively small portion of the chain, namely the inter-domain linker and side chains undergo significant conformational changes upon domain rearrangement.

Hinges by Secondary Structures

A study by Hayward found that the termini of α-helices and β-sheets form hinges in a large number of cases. Many hinges were found to involve two secondary structure elements acting like hinges of a door, allowing an opening and closing motion to occur. This can arise when two neighbouring strands within a β-sheet situated in one domain, diverge apart as they join the other domain. The two resulting termini then form the bending regions between the two domains. α-helices that preserve their hydrogen bonding network when bent are found to behave as mechanical hinges, storing `elastic energy' that drives the closure of domains for rapid capture of a substrate.

Helical to Extended Conformation

The interconversion of helical and extended conformations at the site of a domain boundary is not uncommon. In calmodulin, torsion angles change for five residues in the middle of a domain linking α-helix. The helix is split into two, almost perpendicular, smaller helices separated by four residues of an extended strand.

Shear Motions

Shear motions involve a small sliding movement of domain interfaces, controlled by the amino acid side chains within the interface. Proteins displaying shear motions often have a layered architecture: stacking of secondary structures. The interdomain linker has merely the role of keeping the domains in close proximity.

Domain Motion and Functional Dynamics in Enzymes

The analysis of the internal dynamics of structurally different, but functionally similar enzymes has highlighted a common relationship between the positioning of the active site and the two principal protein sub-domains. In fact, for several members of the hydrolase superfamily, the catalytic site is located close to the interface separating the two principal quasi-rigid domains. Such positioning appears instrumental for maintaining the precise geometry of the active site, while allowing for an appreciable functionally oriented modulation of the flanking regions resulting from the relative motion of the two sub-domains.

Implications for Macromolecular Evolution

Evidence suggests that protein dynamics are important for function, e.g. enzyme catalysis in DHFR, yet they are also posited to facilitate the acquisition of new functions by molecular evolution. This argument suggests that proteins have evolved to have stable, mostly unique folded structures, but the unavoidable residual flexibility leads to some degree of functional promiscuity, which can be amplified/harnessed/diverted by subsequent mutations. However, there is growing awareness that intrinsically unstructured proteins are quite prevalent in eukaryotic genomes,casting further doubt on the simplest interpretation of Anfinsen's dogma: "sequence determines structure (singular)". In effect, the new paradigm is characterized by the addition of two caveats: "sequence and cellular environment determine structural ensemble".

REFERENCES

1. Wrighton, Mark S.; Ebbing, Darrell D. (1993). General chemistry (4th ed.). Boston: Houghton Mifflin. ISBN 0-395-63696-5.
2. Danson, Michael; Eisenthal, Robert (2002). Enzyme assays: a practical approach. Oxford [Oxfordshire]: Oxford University Press. ISBN 0-19-963820-9.
3. Xie XS, Lu HP (June 1999). "Single-molecule enzymology". J. Biol. Chem.

274 (23): 15967–70. doi:10.1074/jbc.274.23.15967. PMID 10347141.

4. Lu H (2004). "Single-molecule spectroscopy studies of conformational change dynamics in enzymatic reactions". Current pharmaceutical biotechnology. 5 (3): 261–9. doi:10.2174/1389201043376887. PMID 15180547.
5. Schnell J, Dyson H, Wright P (2004). "Structure, dynamics, and catalytic function of dihydrofolate reductase". Annual Review of Biophysics and Biomolecular Structure. 33: 119–40. doi:10.1146/annurev.biophys.33.110502.133613. PMID 15139807.
6. Gibson QH (1969). "Rapid mixing: Stopped flow". Methods in Enzymology. Methods in Enzymology. 16: 187–228. doi:10.1016/S0076-6879(69)16009-7. ISBN 978-0-12-181873-9.
7. Duggleby RG (1995). "Analysis of enzyme progress curves by non-linear regression". Methods in Enzymology. Methods in Enzymology. 249: 61–90. doi:10.1016/0076-6879(95)49031-0. ISBN 978-0-12-182150-0. PMID 7791628.
8. Murray JB, Dunham CM, Scott WG (January 2002). "A pH-dependent conformational change, rather than the chemical step, appears to be rate-limiting in the hammerhead ribozyme cleavage reaction". J. Mol. Biol. 315 (2): 121–30. doi:10.1006/jmbi.2001.5145. PMID 11779233.
9. Michaelis L. and Menten M.L. Kinetik der Invertinwirkung Biochem. Z. 1913; 49:333–369 English translation Accessed 6 April 2007
10. Stroppolo ME, Falconi M, Caccuri AM, Desideri A (2001). "Superefficient enzymes". Cell. Mol. Life Sci. 58 (10): 1451–60. doi:10.1007/PL00000788. PMID 11693526.
11. Walsh R, Martin E, Darvesh S (Jan 2010). "A method to describe enzyme-catalyzed reactions by combining steady state and time course enzyme kinetic parameters". Biochim Biophys Acta. 1800 (1): 1–5. doi:10.1016/j.bbagen.2009.10.007. PMID 19840832.
12. Beal, S. L. (1983). "Computation of the explicit solution to the Michaelis-Menten equation". Journal of Pharmacokinetics and Biopharmaceutics. 11 (6): 641–657. doi:10.1007/BF01059062. PMID 6689584.
13. Schnell, S.; Mendoza, C. (1997). "Closed Form Solution for Time-dependent Enzyme Kinetics". Journal of Theoretical Biology. 187 (2): 207–212. doi:10.1006/jtbi.1997.0425.
14. Goudar, CT; Sonnad, JR; Duggleby, RG (1999). "Parameter estimation using a direct solution of the integrated Michaelis-Menten equation". Biochimica et Biophysica Acta. 1429 (2): 377–83. doi:10.1016/s0167-4838(98)00247-7. PMID 9989222. Archived from the original (PDF) on 2015-11-09.
15. Goudar, C. T.; Harris, S. K.; McInerney, M. J.; Suflita, J. M. (2004). "Progress curve analysis for enzyme and microbial kinetic reactions using explicit solutions based on the Lambert W function". Journal of Microbiological Methods. 59 (3): 317–326. doi:10.1016/j.mimet.2004.06.013. PMID 15488275.
16. Jones ME (1992). "Analysis of algebraic weighted least-squares estimators for

enzyme parameters". Biochem. J. 288 (Pt 2): 533–8. PMC 1132043 . PMID 1463456.

17. Tseng SJ, Hsu JP (August 1990). "A comparison of the parameter estimating procedures for the Michaelis-Menten model". J. Theor. Biol. 145 (4): 457–64. doi:10.1016/S0022-5193(05)80481-3. PMID 2246896.
18. Bravo IG, Busto F, De Arriaga D, et al. (September 2001). "A normalized plot as a novel and time-saving tool in complex enzyme kinetic analysis". Biochem. J. 358 (Pt 3): 573–83. PMC 1222113 . PMID 11577687.
19. Almaas E, Kovács B, Vicsek T, Oltvai ZN, Barabási AL (February 2004). "Global organization of metabolic fluxes in the bacterium Escherichia coli". Nature. 427 (6977): 839–43. Bibcode:2004Natur.427..839A. doi:10.1038/nature02289. PMID 14985762.
20. Reed JL, Vo TD, Schilling CH, Palsson BO (2003). "An expanded genome-scale model of Escherichia coli K-12 (iJR904 GSM/GPR)". Genome Biol. 4 (9): R54. doi:10.1186/gb-2003-4-9-r54. PMC 193654 . PMID 12952533.
21. for a complete derivation, see here
22. Dirr H, Reinemer P, Huber R (1994). "X-ray crystal structures of cytosolic glutathione S-transferases. Implications for protein architecture, substrate recognition and catalytic function". Eur. J. Biochem. 220 (3): 645–61. doi:10.1111/j.1432-1033.1994.tb18666.x. PMID 8143720.
23. Stone SR, Morrison JF (July 1988). "Dihydrofolate reductase from Escherichia coli: the kinetic mechanism with NADPH and reduced acetylpyridine adenine dinucleotide phosphate as substrates". Biochemistry. 27 (15): 5493–9. doi:10.1021/bi00415a016. PMID 3052577.
24. Fisher PA (1994). "Enzymologic mechanism of replicative DNA polymerases in higher eukaryotes". Progress in Nucleic Acid Research and Molecular Biology. Progress in Nucleic Acid Research and Molecular Biology. 47: 371–97. doi:10.1016/S0079-6603(08)60257-3. ISBN 978-0-12-540047-3. PMID 8016325.
25. Akerman SE, Müller S (2003). "2-Cys peroxiredoxin PfTrx-Px1 is involved in the antioxidant defence of Plasmodium falciparum". Mol. Biochem. Parasitol. 130 (2): 75–81. doi:10.1016/S0166-6851(03)00161-0. PMID 12946843.
26. Bravo IG, Barrallo S, Ferrero MA, Rodríguez-Aparicio LB, Martínez-Blanco H, Reglero A (September 2001). "Kinetic properties of the acylneuraminate cytidylyltransferase from Pasteurella haemolytica A2". Biochem. J. 358 (Pt 3): 585–98. PMC 1222114 . PMID 11577688.
27. Kraut J (1977). "Serine proteases: structure and mechanism of catalysis". Annu. Rev. Biochem. 46: 331–58. doi:10.1146/annurev.bi.46.070177.001555. PMID 332063.
28. Ricard J, Cornish-Bowden A (July 1987). "Co-operative and allosteric enzymes: 20 years on". Eur. J. Biochem. 166 (2): 255–72. doi:10.1111/j.1432-1033.1987.tb13510.x. PMID 3301336.
29. Ward WH, Fersht AR (July 1988). "Tyrosyl-tRNA synthetase acts as an asymmetric

dimer in charging tRNA. A rationale for half-of-the-sites activity". Biochemistry. 27 (15): 5525–30. doi:10.1021/bi00415a021. PMID 3179266.

30. Helmstaedt K, Krappmann S, Braus GH (September 2001). "Allosteric Regulation of Catalytic Activity: Escherichia coli Aspartate Transcarbamoylase versus Yeast Chorismate Mutase". Microbiol. Mol. Biol. Rev. 65 (3): 404–21, table of contents. doi:10.1128/MMBR.65.3.404-421.2001. PMC 99034 . PMID 11528003.

31. Schirmer T, Evans PR (January 1990). "Structural basis of the allosteric behaviour of phosphofructokinase". Nature. 343 (6254): 140–5. Bibcode:1990Natur.343..140S. doi:10.1038/343140a0. PMID 2136935.

32. Hill, A. V. The possible effects of the aggregation of the molecules of haemoglobin on its dissociation curves. J. Physiol. (Lond.), 1910 40, iv–vii.

33. Hartley BS, Kilby BA (February 1954). "The reaction of p-nitrophenyl esters with chymotrypsin and insulin". Biochem. J. 56 (2): 288–97. PMC 1269615 . PMID 13140189.

34. Fersht, Alan (1999). Structure and mechanism in protein science: a guide to enzyme catalysis and protein folding. San Francisco: W.H. Freeman. ISBN 0-7167-3268-8.

35. Bender ML, Begué-Cantón ML, Blakeley RL, et al. (December 1966). "The determination of the concentration of hydrolytic enzyme solutions: alpha-chymotrypsin, trypsin, papain, elastase, subtilisin, and acetylcholinesterase". J. Am. Chem. Soc. 88 (24): 5890–913. doi:10.1021/ja00976a034. PMID 5980876.

36. Cleland WW (January 2005). "The use of isotope effects to determine enzyme mechanisms". Arch. Biochem. Biophys. 433 (1): 2–12. doi:10.1016/j.abb.2004.08.027. PMID 15581561.

37. Northrop D (1981). "The expression of isotope effects on enzyme-catalyzed reactions". Annu. Rev. Biochem. 50: 103–31. doi:10.1146/annurev.bi.50.070181.000535. PMID 7023356.

38. Baillie T, Rettenmeier A (1986). "Drug biotransformation: mechanistic studies with stable isotopes". Journal of clinical pharmacology. 26 (6): 448–51. doi:10.1002/j.1552-4604.1986.tb03556.x. PMID 3734135.

39. Cleland WW (1982). "Use of isotope effects to elucidate enzyme mechanisms". CRC Crit. Rev. Biochem. 13 (4): 385–428. doi:10.3109/10409238209108715. PMID 6759038.

40. Christianson DW, Cox JD (1999). "Catalysis by metal-activated hydroxide in zinc and manganese metalloenzymes". Annu. Rev. Biochem. 68: 33–57. doi:10.1146/annurev.biochem.68.1.33. PMID 10872443.

41. Kraut D, Carroll K, Herschlag D (2003). "Challenges in enzyme mechanism and energetics". Annu. Rev. Biochem. 72: 517–71. doi:10.1146/annurev.biochem.72.121801.161617. PMID 12704087.

42. Walsh, R.; Martin, E.; Darvesh, S. (2011). "Limitations of conventional inhibitor classifications". Integrative Biology. Royal Society of Chemistry. 3 (12): 1197–1201. doi:10.1039/c1ib00053e. PMID 22038120.

43. Walsh, R.; Martin, E.; Darvesh, S. (2007). "A versatile equation to describe reversible enzyme inhibition and activation kinetics: Modeling β-galactosidase and butyrylcholinesterase". Biochimica et Biophysica Acta (BBA) - General Subjects. 1770 (5): 733–746. doi:10.1016/j.bbagen.2007.01.001. PMID 17307293.

44. Walsh, Ryan (2012). "Ch. 17. Alternative Perspectives of Enzyme Kinetic Modeling". In Ekinci, Deniz. Medicinal Chemistry and Drug Design (PDF). InTech. pp. 357–371. ISBN 978-953-51-0513-8.

45. Koshland DE (February 1958). "Application of a Theory of Enzyme Specificity to Protein Synthesis". Proc. Natl. Acad. Sci. U.S.A. 44 (2): 98–104. Bibcode:1958PNAS...44...98K. doi:10.1073/pnas.44.2.98. PMC 335371 . PMID 16590179.

46. Hammes G (2002). "Multiple conformational changes in enzyme catalysis". Biochemistry. 41 (26): 8221–8. doi:10.1021/bi0260839. PMID 12081470.

47. Sutcliffe M, Scrutton N (2002). "A new conceptual framework for enzyme catalysis. Hydrogen tunnelling coupled to enzyme dynamics in flavoprotein and quinoprotein enzymes". Eur. J. Biochem. 269 (13): 3096–102. doi:10.1046/j.1432-1033.2002.03020.x. PMID 12084049.

48. Henri V (1902). "Theorie generale de l'action de quelques diastases". Compt. Rend. Hebd. Acad. Sci. Paris. 135: 916–9.

49. Sørensen PL (1909). "Enzymstudien {II}. Über die Messung und Bedeutung der Wasserstoffionenkonzentration bei enzymatischen Prozessen [Enzyme studies III: About the measurement and significance of the hydrogen ion concentration in enzymatic processes]. Biochem. Z. (in German). 21: 131–304.

50. Michaelis L, Menten M (1913). Die Kinetik der Invertinwirkung [The Kinetics of Invertase Action]. Biochem. Z. (in German). 49: 333–369.; Michaelis L, Menten ML, Johnson KA, Goody RS (2011). The original Michaelis constant: translation of the 1913 Michaelis-Menten paper. Biochemistry. 50 (39): 8264–9. doi:10.1021/bi201284u. PMC 3381512 . PMID 21888353.

51. Briggs GE, Haldane JB (1925). A Note on the Kinetics of Enzyme Action. The Biochemical Journal. 19 (2): 339–339. doi:10.1042/bj0190338. PMC 1259181 . PMID 16743508.

52. Flomenbom O, Velonia K, Loos D, Masuo S, Cotlet M, Engelborghs Y, Hofkens J, Rowan AE, Nolte RJ, Van der Auweraer M, de Schryver FC, Klafter J (Feb 2005). Stretched exponential decay and correlations in the catalytic activity of fluctuating single lipase molecules. Proceedings of the National Academy of Sciences of the United States of America. 102 (7): 2368–2372. Bibcode:2005PNAS..102.2368F. doi:10.1073/pnas.0409039102. PMC 548972 . PMID 15695587.

53. English BP, Min W, van Oijen AM, Lee KT, Luo G, Sun H, Cherayil BJ, Kou SC, Xie XS (Feb 2006). Ever-fluctuating single enzyme molecules: Michaelis-Menten equation revisited. Nature Chemical Biology. 2 (2): 87–94. doi:10.1038/nchembio759. PMID 16415859.

54. Lu HP, Xun L, Xie XS (Dec 1998). Single-molecule enzymatic dynamics. Science. 282 (5395): 1877–1882. doi:10.1126/science.282.5395.1877. PMID 9836635.

55. Xue X, Liu F, Ou-Yang ZC (Sep 2006). Single molecule Michaelis-Menten equation beyond quasistatic disorder. Physical Review E. 74 (3 Pt 1): 030902. Bibcode:2006PhRvE..74c0902X. doi:10.1103/PhysRevE.74.030902. PMID 17025584.

56. Bevc S.; Konc J.; Stojan J.; Hodoščеk M.; Penca M.; Matej Praprotnik M.; Janežič D. (2011). ENZO: A Web Tool for Derivation and Evaluation of Kinetic Models of Enzyme Catalyzed Reactions. PLoS ONE. 6 (7): e22265. doi:10.1371/journal.pone.0022265. PMC 3139599 . PMID 21818304. ENZO server

57. Bu Z, Callaway DJ (2011). Proteins MOVE! Protein dynamics and long-range allostery in cell signaling. Advances in Protein Chemistry and Structural Biology. Advances in Protein Chemistry and Structural Biology. 83: 163–221. doi:10.1016/B978-0-12-381262-9.00005-7. ISBN 9780123812629. PMID 21570668.

58. J. Monod, J. Wyman, J.P. Changeux. (1965). On the nature of allosteric transitions:A plausible model. J. Mol. Biol., May;12:88-118.

59. D.E. Jr Koshland, G. Némethy, D. Filmer (1966) Comparison of experimental binding data and theoretical models in proteins containing subunits. Biochemistry. Jan;5(1):365-8

60. E. K. Jaffe (2005). Morpheeins - a new structural paradigm for allosteric regulation. Trends Biochem. Sci. 30 (9): 490–497. doi:10.1016/j.tibs.2005.07.003. PMID 16023348.

61. Hilser, V. J., Wrabl, J. O. & Motlagh, H. N. (2012) Structural and energetic basis of allostery. Ann. Rev. Biophys. 41, 585–609

62. LeVine, M.V., and Weinstein, H. (2015) AIM for Allostery: Using the Ising Model to Understand Information Processing and Transmission in Allosteric Biomolecular Systems. Entropy. 17(5), 2895-2918

63. Cuendet, M.A., Weinstein, H., and LeVine, M.V. (2016) The allostery landscape: quantifying thermodynamic coupling in biomolecular systems. Journal of Chemical Theory and Computation.

64. Huang, Z.; Zhu, L.; Cao, Y.; Wu, G.; Liu, X.; et al. (2011). ASD: a comprehensive database of allosteric proteins and modulators. Nucleic Acids Res. 39: D663–669. doi:10.1093/nar/gkq1022.

65. Panjkovich, Alejandro; Daura, Xavier (2012-01-01). Exploiting protein flexibility to predict the location of allosteric sites. BMC Bioinformatics. 13: 273. doi:10.1186/1471-2105-13-273. ISSN 1471-2105. PMC 3562710 . PMID 23095452.

66. Süel, Gürol M.; Lockless, Steve W.; Wall, Mark A.; Ranganathan, Rama (2003-01-01). Evolutionarily conserved networks of residues mediate allosteric communication in proteins. Nature Structural Biology. 10 (1): 59–69. doi:10.1038/nsb881. ISSN 1072-8368. PMID 12483203.

67. Mitternacht, Simon; Berezovsky, Igor N. (2011-09-01). Binding leverage as a molecular basis for allosteric regulation. PLOS Computational Biology. 7 (9): e1002148. doi:10.1371/journal.pcbi.1002148. ISSN 1553-7358. PMC 3174156 . PMID 21935347.

68. Gasper, Paul M.; Fuglestad, Brian; Komives, Elizabeth A.; Markwick, Phineus R. L.; McCammon, J. Andrew (2012-12-26). Allosteric networks in thrombin distinguish procoagulant vs. anticoagulant activities. Proceedings of the National Academy of Sciences of the United States of America. 109 (52): 21216–21222. doi:10.1073/pnas.1218414109. ISSN 1091-6490. PMC 3535651 . PMID 23197839.

69. Ghosh, Amit; Vishveshwara, Saraswathi (2008-11-04). Variations in clique and community patterns in protein structures during allosteric communication: investigation of dynamically equilibrated structures of methionyl tRNA synthetase complexes. Biochemistry. 47 (44): 11398–11407. doi:10.1021/bi8007559. ISSN 1520-4995. PMID 18842003.

70. Sethi, Anurag; Eargle, John; Black, Alexis A.; Luthey-Schulten, Zaida (2009-04-21). Dynamical networks in tRNA:protein complexes. Proceedings of the National Academy of Sciences of the United States of America. 106 (16): 6620–6625. doi:10.1073/pnas.0810961106. ISSN 1091-6490. PMC 2672494 . PMID 19351898.

71. Vanwart, Adam T.; Eargle, John; Luthey-Schulten, Zaida; Amaro, Rommie E. (2012-08-14). Exploring residue component contributions to dynamical network models of allostery. Journal of Chemical Theory and Computation. 8 (8): 2949–2961. doi:10.1021/ct300377a. ISSN 1549-9626. PMC 3489502 . PMID 23139645.

72. Rivalta, Ivan; Sultan, Mohammad M.; Lee, Ning-Shiuan; Manley, Gregory A.; Loria, J. Patrick; Batista, Victor S. (2012-05-29). Allosteric pathways in imidazole glycerol phosphate synthase. Proceedings of the National Academy of Sciences of the United States of America. 109 (22): E1428–1436. doi:10.1073/pnas.1120536109. ISSN 1091-6490. PMC 3365145 . PMID 22586084.

73. Clarke, Declan; Sethi, Anurag; Li, Shantao; Kumar, Sushant; Chang, Richard W. F.; Chen, Jieming; Gerstein, Mark (2016-04-02). Identifying Allosteric Hotspots with Dynamics: Application to Inter- and Intra-species Conservation. Structure (London, England: 1993). 24: 826–37. doi:10.1016/j.str.2016.03.008. ISSN 1878-4186. PMID 27066750.

74. Srinivasan B; et al. (2014). Allosteric regulation and substrate activation in cytosolic nucleotidase II from Legionella pneumophila.. FEBS J. 281 (6): 1613–1628. doi:10.1111/febs.12727. PMC 3982195 . PMID 24456211.

75. Edelstein, SJ (1975). Cooperative interactions of hemoglobin. Annu Rev Biochem. 44: 209–232. doi:10.1146/annurev.bi.44.070175.001233. PMID 237460.

76. A. Christopoulos, L.T. May, V.A. Avlani and P.M. Sexton (2004) G protein-coupled receptor allosterism:the promise and the problem(s). Biochemical Society Transactions Volume 32, part 5

77. May, L.T.; Leach, K.; Sexton, P.M.; Christopoulos, A. (2007). Allosteric Modulation of G Protein–Coupled Receptors. Annu. Rev. Pharmacol. Toxicol. 47: 1–51. doi:10.1146/annurev.pharmtox.47.120505.105159. PMID 17009927.

78. E. K. Jaffe (2010). Morpheeins - A new pathway for allosteric drug discovery. Open Conf. Proc. J. 1: 1–6. doi:10.2174/2210289201001010001. PMC 3107518

. PMID 21643557.

79. Bu Z, Callaway DJ (2011). Proteins move! Protein dynamics and long-range allostery in cell signaling. Advances in Protein Chemistry and Structural Biology. 83: 163–221. doi:10.1016/B978-0-12-381262-9.00005-7. PMID 21570668.

80. Fraser JS, Clarkson MW, Degnan SC, Erion R, Kern D, Alber T (Dec 2009). Hidden alternative structures of proline isomerase essential for catalysis. Nature. 462 (7273): 669–673. Bibcode:2009Natur.462..669F. doi:10.1038/nature08615. PMID 19956261.

81. Frauenfelder H, Sligar SG, Wolynes PG (Dec 1991). The energy landscapes and motions of proteins. Science. 254 (5038): 1598–1603. Bibcode:1991Sci...254.1598F. doi:10.1126/science.1749933. PMID 1749933.

82. Davis IW, Arendall WB, Richardson DC, Richardson JS (Feb 2006). The backrub motion: how protein backbone shrugs when a sidechain dances. Structure (London, England : 1993). 14 (2): 265–274. doi:10.1016/j.str.2005.10.007. PMID 16472746.

83. Fraser JS, van den Bedem H, Samelson AJ, Lang PT, Holton JM, Echols N, Alber T (Sep 2011). Accessing protein conformational ensembles using room-temperature X-ray crystallography. Proceedings of the National Academy of Sciences of the United States of America. 108 (39): 16247–16252. doi:10.1073/pnas.1111325108. PMID 21918110.

84. Bu Z, Cook J, Callaway DJ (Sep 2001). Dynamic regimes and correlated structural dynamics in native and denatured alpha-lactalbumin. Journal of Molecular Biology. 312 (4): 865–873. doi:10.1006/jmbi.2001.5006. PMID 11575938.

85. Farago B, Li J, Cornilescu G, Callaway DJ, Bu Z (Nov 2010). Activation of nanoscale allosteric protein domain motion revealed by neutron spin echo spectroscopy. Biophysical Journal. 99 (10): 3473–3482. Bibcode:2010BpJ....99.3473F. doi:10.1016/j.bpj.2010.09.058. PMC 2980739 . PMID 21081097.

86. Bu Z, Biehl R, Monkenbusch M, Richter D, Callaway DJ (Dec 2005). Coupled protein domain motion in Taq polymerase revealed by neutron spin-echo spectroscopy. Proceedings of the National Academy of Sciences of the United States of America. 102 (49): 17646–17651. Bibcode:2005PNAS..10217646B. doi:10.1073/pnas.0503388102. PMC 1345721 . PMID 16306270.

87. Potestio R, Pontiggia F, Micheletti C (Jun 2009). Coarse-grained description of protein internal dynamics: an optimal strategy for decomposing proteins in rigid subunits. Biophysical Journal. 96 (12): 4993–5002. Bibcode:2009BpJ....96.4993P. doi:10.1016/j.bpj.2009.03.051. PMC 2712024 . PMID 19527659.

88. Baron R, Vellore NA (Jul 2012). LSD1/CoREST is an allosteric nanoscale clamp regulated by H3-histone-tail molecular recognition. Proceedings of the National Academy of Sciences of the United States of America. 109 (31): 12509–14. Bibcode:2012PNAS..10912509B. doi:10.1073/pnas.1207892109. PMC 3411975 . PMID 22802671.

89. Gerstein M, Lesk AM, Chothia C (Jun 1994). Structural mechanisms for domain movements in proteins. Biochemistry. 33 (22): 6739–49. doi:10.1021/bi00188a001. PMID 8204609.

90. Kamerlin SC, Warshel A (May 2010). At the dawn of the 21st century: Is dynamics the missing link for understanding enzyme catalysis?. Proteins. 78 (6): 1339–75. doi:10.1002/prot.22654. PMC 2841229 . PMID 20099310.

91. Herzberg O, Chen CC, Kapadia G, McGuire M, Carroll LJ, Noh SJ, Dunaway-Mariano D (Apr 1996). Swiveling-domain mechanism for enzymatic phosphotransfer between remote reaction sites. Proceedings of the National Academy of Sciences of the United States of America. 93 (7): 2652–7. Bibcode:1996PNAS...93.2652H. doi:10.1073/pnas.93.7.2652. PMC 39685 . PMID 8610096.

92. Janin J, Wodak SJ (1983). Structural domains in proteins and their role in the dynamics of protein function. Progress in Biophysics and Molecular Biology. 42 (1): 21–78. doi:10.1016/0079-6107(83)90003-2. PMID 6353481.

93. Hayward S (Sep 1999). Structural principles governing domain motions in proteins. Proteins. 36 (4): 425–35. doi:10.1002/(SICI)1097-0134(19990901)36:4<425::AID-PROT6>3.0.CO;2-S. PMID 10450084.

94. Meador WE, Means AR, Quiocho FA (Aug 1992). Target enzyme recognition by calmodulin: 2.4 A structure of a calmodulin-peptide complex. Science. 257 (5074): 1251–1255. Bibcode:1992Sci...257.1251M. doi:10.1126/science.1519061. PMID 1519061.

95. Ikura M, Clore GM, Gronenborn AM, Zhu G, Klee CB, Bax A (May 1992). Solution structure of a calmodulin-target peptide complex by multidimensional NMR. Science. 256 (5057): 632–638. Bibcode:1992Sci...256..632I. doi:10.1126/science.1585175. PMID 1585175.

96. Tokuriki N, Tawfik DS (Apr 2009). Protein dynamism and evolvability. Science. 324 (5924): 203–207. Bibcode:2009Sci...324..203T. doi:10.1126/science.1169375. PMID 19359577.

97. Dyson HJ, Wright PE (Mar 2005). Intrinsically unstructured proteins and their functions. Nature Reviews Molecular Cell Biology. 6 (3): 197–208. doi:10.1038/nrm1589. PMID 15738986.

CHAPTER 6

Stuctural Determination Techniques

Chapter Contents

- **Introduction**
- **X-Ray Crystallography**
- **Nuclear Magnetic Resonance Spectroscopy**
- **Atomic-Force Microscopy**
- **Small-Angle Scattering**
- **Small-Angle X-Ray Scattering**

Structure determination is a procedure by which the three-dimensional atomic coordinates of a molecule or biomolecule are solved using an analytical technique. Many techniques are used in structure determination, most commonly X-ray crystallography, NMR spectroscopy, electron microscopy and molecular modelling.

INTRODUCTION

An electron microscope is a microscope that uses a beam of accelerated electrons as a source of illumination. As the wavelength of an electron can be up to 100,000 times shorter than that of visible light photons, electron microscopes have a higher resolving power than light microscopes and can reveal the structure of smaller objects. A transmission electron microscope can achieve better than 50 pm resolution and magnifications of up to about 10,000,000x whereas most light microscopes are limited by diffraction to about 200 nm resolution and useful magnifications below 2000x.

Transmission electron microscopes use electrostatic and electromagnetic lenses to control the electron beam and focus it to form an image. These electron optical lenses are analogous to the glass lenses of an optical light microscope. Electron microscopes are used to investigate the ultrastructure of a wide range of biological and inorganic specimens including microorganisms, cells, large molecules, biopsy samples, metals, and crystals. Industrially, electron microscopes are often used for quality control and failure analysis. Modern electron microscopes produce electron micrographs using specialized digital cameras and frame grabbers to capture the image.

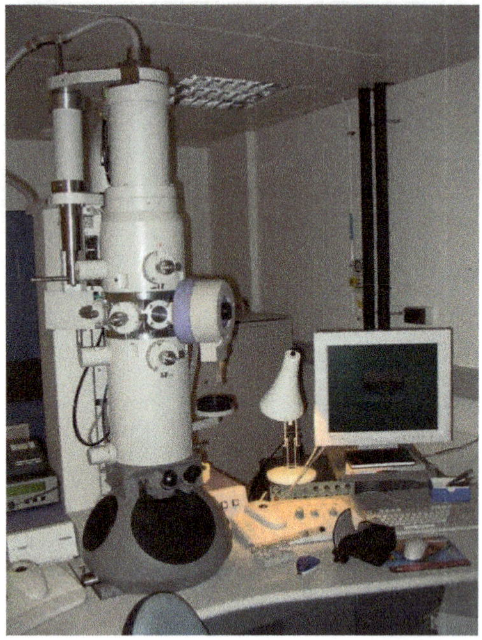

A modern transmission electron microscope

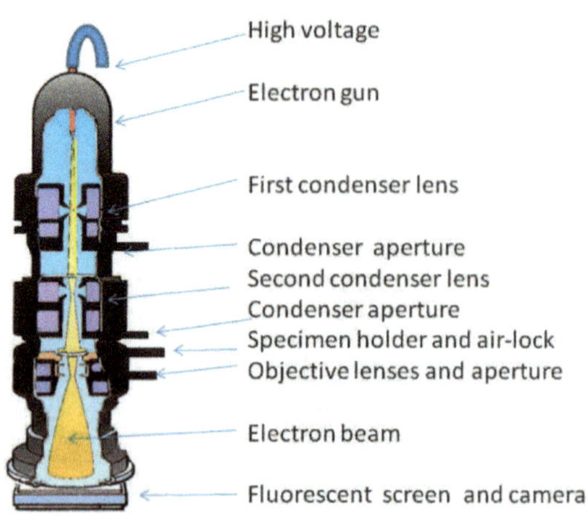

Diagram of a transmission electron microscope

Electron microscope constructed by Ernst Ruska in 1933

The first electromagnetic lens was developed in 1926 by Hans Busch.

According to Dennis Gabor, the physicist Leó Szilárd tried in 1928 to convince Busch to build an electron microscope, for which he had filed a patent.

German physicist Ernst Ruska and the electrical engineer Max Knoll constructed the prototype electron microscope in 1931, capable of four-hundred-power magnification; the apparatus was the first demonstration of the principles of electron microscopy.Two years later, in 1933, Ruska built an electron microscope that exceeded the resolution attainable with an optical (light) microscope. Moreover, Reinhold Rudenberg, the scientific director of Siemens-Schuckertwerke, obtained the patent for the electron microscope in May 1931.

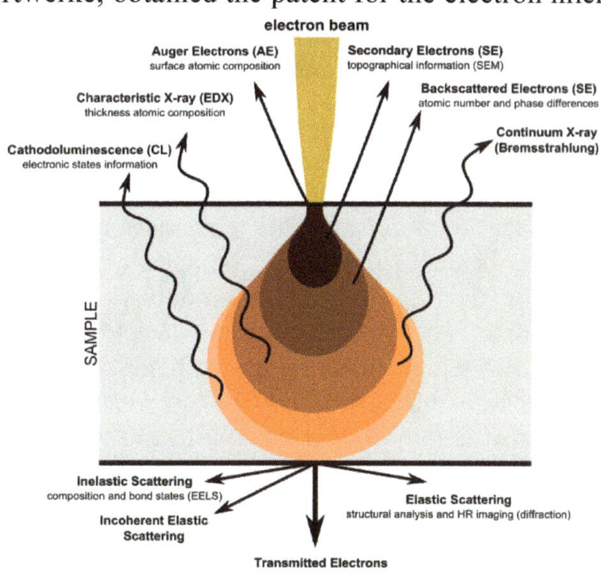

Diagram illustrating the phenomena resulting of the interaction of highly energetic electrons with matter.

In 1932, Ernst Lubcke of Siemens & Halske built and obtained images from a prototype electron microscope, applying concepts described in the Rudenberg patent applications. Five years later (1937), the firm financed the work of Ernst Ruska and Bodo von Borries, and employed Helmut Ruska (Ernst's brother) to develop applications for the microscope, especially with biological specimens. Also in 1937, Manfred von Ardenne pioneered the scanning electron microscope.The first commercial electron microscope was produced in 1938 by Siemens. The first North American electron microscope was constructed in 1938, at the University of Toronto, by Eli Franklin Burton and students Cecil Hall, James Hillier, and Albert Prebus; and Siemens produced a transmission electron microscope (TEM) in 1939.Although contemporary transmission electron microscopes are capable of two million-power magnification, as scientific instruments, they remain based upon Ruska's prototype.

Types

Transmission Electron Microscope (TEM)

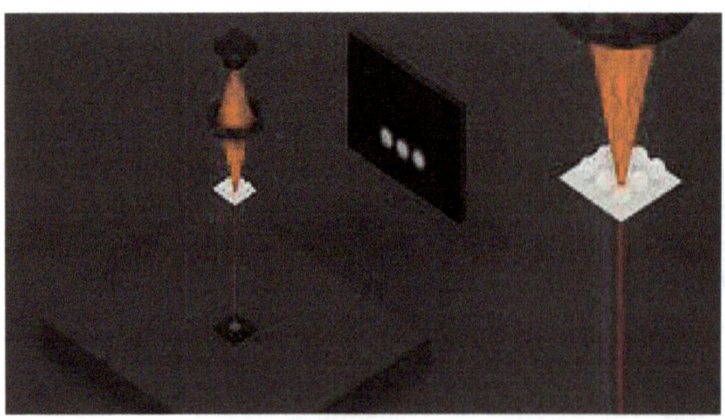

Operating principe of a Transmission Electron Microscope

The original form of electron microscope, the transmission electron microscope (TEM) uses a high voltage electron beam to illuminate the specimen and create an image. The electron beam is produced by an electron gun, commonly fitted with a tungsten filament cathode as the electron source. The electron beam is accelerated by an anode typically at +100 keV (40 to 400 keV) with respect to the cathode, focused by electrostatic and electromagnetic lenses, and transmitted through the specimen that is in part transparent to electrons and in part scatters them out of the beam. When it emerges from the specimen, the electron beam carries information about the structure of the specimen that is magnified by the objective lens system of the microscope. The spatial variation in this information (the "image") may be viewed by projecting the magnified electron image onto a fluorescent viewing screen coated with a phosphor or scintillator material such as zinc sulfide. Alternatively, the image can be photographically recorded by exposing a photographic film or plate directly to the electron beam, or a high-resolution phosphor may be coupled by means of a lens optical system or a fibre optic light-guide to the sensor of a digital camera. The image detected by the digital camera may be displayed on a monitor or computer.

The resolution of TEMs is limited primarily by spherical aberration, but a new generation of aberration correctors have been able to partially overcome spherical aberration to increase resolution. Hardware correction of spherical aberration for the high-resolution transmission electron microscopy (HRTEM) has allowed the production of images with resolution below 0.5 angstrom (50 picometres) and magnifications above 50 million times. The ability to determine the positions of atoms within materials has made the HRTEM an important tool for nano-technologies research and development.

Transmission electron microscopes are often used in electron diffraction mode. The advantages of electron diffraction over X-ray crystallography are that the specimen need not be a single crystal or even a polycrystalline powder, and also that the Fourier transform reconstruction of the object's magnified structure occurs physically and thus avoids the need for solving the phase problem faced by the X-ray crystallographers after obtaining their X-ray diffraction patterns of a single crystal or polycrystalline powder.

The major disadvantage of the transmission electron microscope is the need for extremely thin sections of the specimens, typically about 100 nanometers. Biological specimens are typically required to be chemically fixed, dehydrated and embedded in a polymer resin to stabilize them sufficiently to allow ultrathin sectioning. Sections of biological specimens, organic polymers and similar materials may require special treatment with heavy atom labels in order to achieve the required image contrast.

Scanning Electron Microscope (SEM)

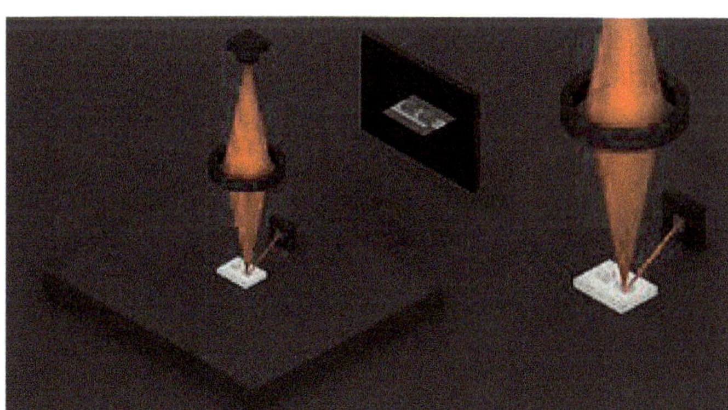

Operating principe of a Scanning Electron Microscope

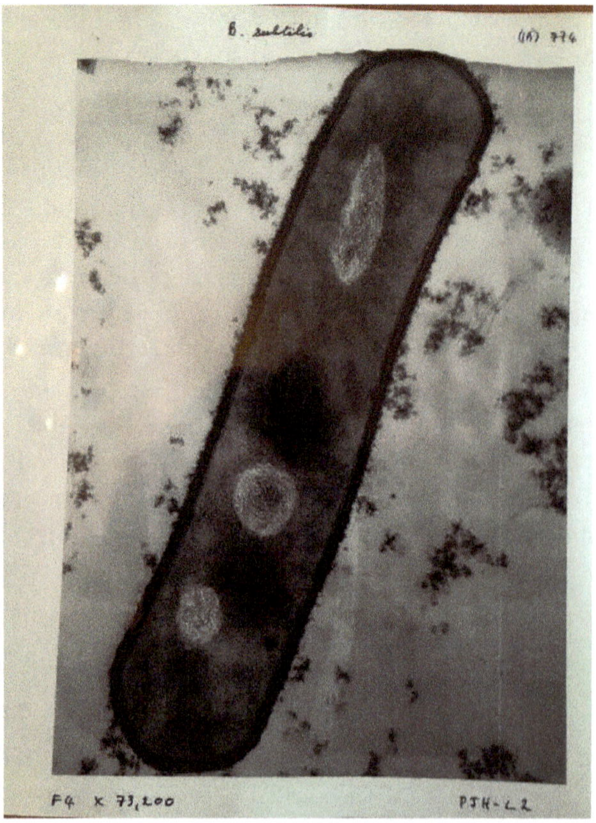

Image of bacillus subtilis taken with a 1960s electron microscope.

The SEM produces images by probing the specimen with a focused electron beam that is scanned across a rectangular area of the specimen (raster scanning). When the electron beam interacts with the specimen, it loses energy by a variety of mechanisms. The lost energy is converted into alternative forms such as heat, emission of low-energy secondary electrons and high-energy backscattered electrons, light emission (cathodoluminescence) or X-ray emission, all of which provide signals carrying information about the properties of the specimen surface, such as its topography and composition. The image displayed by an SEM maps the varying intensity of any of these signals into the image in a position corresponding to the position of the beam on the specimen when the signal was generated. In the SEM image of an ant shown below and to the right, the image was constructed from signals produced by a secondary electron detector, the normal or conventional imaging mode in most SEMs.

Generally, the image resolution of an SEM is at least an order of magnitude poorer than that of a TEM. However, because the SEM image relies on surface processes rather than transmission, it is able to image bulk samples up to many centimeters in size and (depending on instrument design and settings) has a great depth of field, and so can produce images that are good representations of the three-dimensional shape of the sample. Another advantage of SEM is its variety called environmental scanning electron microscope (ESEM) can produce images of sufficient quality and resolution with the samples being wet or contained in low vacuum or gas. This greatly facilitates imaging biological samples that are unstable in the high vacuum of conventional electron microscopes.

An image of an ant in a scanning electron microscope

Color

In their most common configurations, electron microscopes produce images with a single brightness value per pixel, with the results usually rendered in grayscale. However, often these images are then colorized through the use of feature-detection software, or simply by hand-editing using a graphics editor. This may be done to clarify structure or for aesthetic effect and generally does not add new information about the specimen.

In some configurations information about several specimen properties is gathered per pixel, usually by the use of multiple detectors. In SEM, the attributes of topography and material contrast can be obtained by a pair of backscattered electron detectors and such attributes can be superimposed in a single color image by assigning a different primary color to each attribute. Similarly, a combination of backscattered and secondary electron signals can be assigned to different colors and superimposed on a single color micrograph displaying simultaneously the properties of the specimen.

Some types of detectors used in SEM have analytical capabilities, and can provide several items of data at each pixel. Examples are the Energy-dispersive X-ray spectroscopy (EDS) detectors used in elemental analysis and Cathodoluminescence microscope (CL) systems that analyse the intensity and spectrum of electron-induced luminescence in (for example) geological specimens. In SEM systems using these detectors it is common to color code the signals and superimpose them in a single color image, so that differences in the distribution of the various components of the specimen can be seen clearly and compared. Optionally, the standard secondary electron image can be merged with the one or more compositional channels, so that the specimen's structure and composition can be compared. Such images can be made while maintaining the full integrity of the original signal, which is not modified in any way.

Reflection Electron Microscope (REM)

In the reflection electron microscope (REM) as in the TEM, an electron beam is incident on a surface but instead of using the transmission (TEM) or secondary electrons (SEM), the reflected beam of elastically scattered electrons is detected. This technique is typically coupled with reflection high energy electron diffraction (RHEED) and reflection high-energy loss spectroscopy (RHELS). Another variation is spin-polarized low-energy electron microscopy (SPLEEM), which is used for looking at the microstructure of magnetic domains.

Scanning Transmission Electron Microscope (STEM)

The STEM rasters a focused incident probe across a specimen that (as with the TEM) has been thinned to facilitate detection of electrons scattered through the specimen. The high resolution of the TEM is thus possible in STEM. The focusing action (and aberrations) occur before the electrons hit the specimen in the STEM, but afterward in the TEM. The STEMs use of SEM-like beam rastering simplifies annular dark-field imaging, and other analytical techniques, but also means that image data is acquired in serial rather than in parallel fashion. Often TEM can be equipped with the scanning option and then it can function both as TEM and STEM.

Sample Preparation

An insect coated in gold for viewing with a scanning electron microscope

Materials to be viewed under an electron microscope may require processing to produce a suitable sample. The technique required varies depending on the specimen and the analysis required:

- Chemical fixation – for biological specimens aims to stabilize the specimen's mobile macromolecular structure by chemical crosslinking of proteins with aldehydes such as formaldehyde and glutaraldehyde, and lipids with osmium tetroxide.
- Negative stain – suspensions containing nanoparticles or fine biological material (such as viruses and bacteria) are briefly mixed with a dilute solution of an electron-opaque solution such as ammonium molybdate, uranyl acetate (or formate), or phosphotungstic acid. This mixture is applied to a suitably coated EM grid, blotted, then allowed to dry. Viewing of this preparation in the TEM should be carried out without delay for best results. The method is important in microbiology for fast but crude morphological identification, but can also be used as the basis for high resolution 3D reconstruction using EM tomography methodology when carbon films are used for support. Negative staining is also used for observation of nanoparticles.
- Cryofixation – freezing a specimen so rapidly, in liquid ethane, and maintained at liquid nitrogen or even liquid helium temperatures, so that the water forms vitreous (non-crystalline) ice. This preserves the specimen in a snapshot of its solution state. An entire field called cryo-electron microscopy has branched from this technique. With the development of cryo-electron microscopy of vitreous sections (CEMOVIS), it is now possible to observe samples from virtually any biological specimen close to its native state.
- Dehydration – or replacement of water with organic solvents such as ethanol or acetone, followed by critical point drying or infiltration with embedding resins. Also freeze drying.
- Embedding, biological specimens – after dehydration, tissue for observation in the transmission electron microscope is embedded so it can be sectioned ready for viewing. To do this the tissue is passed through a ‹transition solvent› such as propylene oxide (epoxypropane) or acetone and then infiltrated with an epoxy resin such as Araldite, Epon, or Durcupan; tissues may also be embedded directly in water-miscible acrylic resin. After the resin has been polymerized (hardened) the sample is thin sectioned (ultrathin sections) and stained – it is then ready for viewing.
- Embedding, materials – after embedding in resin, the specimen is usually ground and polished to a mirror-like finish using ultra-fine abrasives. The polishing process must be performed carefully to minimize scratches and other polishing artifacts that reduce image quality.
- Metal shadowing – Metal (e.g. platinum) is evaporated from an overhead electrode and applied to the surface of a biological sample at an angle. The surface topography results in variations in the thickness of the metal that are seen as variations in brightness and contrast in the electron microscope image.
- Replication – A surface shadowed with metal (e.g. platinum, or a mixture of

carbon and platinum) at an angle is coated with pure carbon evaporated from carbon electrodes at right angles to the surface. This is followed by removal of the specimen material (e.g. in an acid bath, using enzymes or by mechanical separation) to produce a surface replica that records the surface ultrastructure and can be examined using transmission electron microscopy.

- Sectioning – produces thin slices of specimen, semitransparent to electrons. These can be cut on an ultramicrotome with a diamond knife to produce ultra-thin sections about 60–90 nm thick. Disposable glass knives are also used because they can be made in the lab and are much cheaper.
- Staining – uses heavy metals such as lead, uranium or tungsten to scatter imaging electrons and thus give contrast between different structures, since many (especially biological) materials are nearly "transparent" to electrons (weak phase objects). In biology, specimens can be stained "en bloc" before embedding and also later after sectioning. Typically thin sections are stained for several minutes with an aqueous or alcoholic solution of uranyl acetate followed by aqueous lead citrate.
- Freeze-fracture or freeze-etch – a preparation method particularly useful for examining lipid membranes and their incorporated proteins in "face on" view. The fresh tissue or cell suspension is frozen rapidly (cryofixation), then fractured by breaking or by using a microtome while maintained at liquid nitrogen temperature. The cold fractured surface (sometimes "etched" by increasing the temperature to about −100 °C for several minutes to let some ice sublime) is then shadowed with evaporated platinum or gold at an average angle of 45° in a high vacuum evaporator. A second coat of carbon, evaporated perpendicular to the average surface plane is often performed to improve stability of the replica coating. The specimen is returned to room temperature and pressure, then the extremely fragile "pre-shadowed" metal replica of the fracture surface is released from the underlying biological material by careful chemical digestion with acids, hypochlorite solution or SDS detergent. The still-floating replica is thoroughly washed free from residual chemicals, carefully fished up on fine grids, dried then viewed in the TEM.
- Ion beam milling – thins samples until they are transparent to electrons by firing ions (typically argon) at the surface from an angle and sputtering material from the surface. A subclass of this is focused ion beam milling, where gallium ions are used to produce an electron transparent membrane in a specific region of the sample, for example through a device within a microprocessor. Ion beam milling may also be used for cross-section polishing prior to SEM analysis of materials that are difficult to prepare using mechanical polishing.
- Conductive coating – an ultrathin coating of electrically conducting material, deposited either by high vacuum evaporation or by low vacuum sputter coating of the sample. This is done to prevent the accumulation of static electric fields at the specimen due to the electron irradiation required during imaging. The coating materials include gold, gold/palladium, platinum, tungsten, graphite, etc.
- Earthing – to avoid electrical charge accumulation on a conductive coated

sample, it is usually electrically connected to the metal sample holder. Often an electrically conductive adhesive is used for this purpose.

Disadvantages

Electron microscopes are expensive to build and maintain, but the capital and running costs of confocal light microscope systems now overlaps with those of basic electron microscopes. Microscopes designed to achieve high resolutions must be housed in stable buildings (sometimes underground) with special services such as magnetic field cancelling systems.

The samples largely have to be viewed in vacuum, as the molecules that make up air would scatter the electrons. An exception is the environmental scanning electron microscope, which allows hydrated samples to be viewed in a low-pressure (up to 20 Torr or 2.7 kPa) and/or wet environment.

Scanning electron microscopes operating in conventional high-vacuum mode usually image conductive specimens; therefore non-conductive materials require conductive coating (gold/palladium alloy, carbon, osmium, etc.). The low-voltage mode of modern microscopes makes possible the observation of non-conductive specimens without coating. Non-conductive materials can be imaged also by a variable pressure (or environmental) scanning electron microscope.

Small, stable specimens such as carbon nanotubes, diatom frustules and small mineral crystals (asbestos fibres, for example) require no special treatment before being examined in the electron microscope. Samples of hydrated materials, including almost all biological specimens have to be prepared in various ways to stabilize them, reduce their thickness (ultrathin sectioning) and increase their electron optical contrast (staining). These processes may result in artifacts, but these can usually be identified by comparing the results obtained by using radically different specimen preparation methods. It is generally believed by scientists working in the field that as results from various preparation techniques have been compared and that there is no reason that they should all produce similar artifacts, it is reasonable to believe that electron microscopy features correspond with those of living cells. Since the 1980s, analysis of cryofixed, vitrified specimens has also become increasingly used by scientists, further confirming the validity of this technique.

X-RAY CRYSTALLOGRAPHY

X-ray crystallography is a technique used for determining the atomic and molecular structure of a crystal, in which the crystalline atoms cause a beam of incident X-rays to diffract into many specific directions. By measuring the angles and intensities of these diffracted beams, a crystallographer can produce a three-dimensional picture of the density of electrons within the crystal. From this electron density, the mean positions of the atoms in the crystal can be determined, as well as their chemical bonds, their disorder and various other information.

Since many materials can form crystals—such as salts, metals, minerals, semiconductors, as well as various inorganic, organic and biological molecules—X-ray crystallography has been fundamental in the development of many scientific fields. In its first decades of use, this method determined the size of atoms, the lengths and types of chemical bonds, and the atomic-scale differences among various materials, especially minerals and alloys. The method also revealed the structure and function of many biological molecules, including vitamins, drugs, proteins and nucleic acids such as DNA. X-ray crystallography is still the chief method for characterizing the atomic structure of new materials and in discerning materials that appear similar by other experiments. X-ray crystal structures can also account for unusual electronic or elastic properties of a material, shed light on chemical interactions and processes, or serve as the basis for designing pharmaceuticals against diseases.

In a single-crystal X-ray diffraction measurement, a crystal is mounted on a goniometer. The goniometer is used to position the crystal at selected orientations. The crystal is illuminated with a finely focused monochromatic beam of X-rays, producing a diffraction pattern of regularly spaced spots known as reflections. The two-dimensional images taken at different orientations are converted into a three-dimensional model of the density of electrons within the crystal using the mathematical method of Fourier transforms, combined with chemical data known for the sample. Poor resolution (fuzziness) or even errors may result if the crystals are too small, or not uniform enough in their internal makeup.

X-ray crystallography is related to several other methods for determining atomic structures. Similar diffraction patterns can be produced by scattering electrons or neutrons, which are likewise interpreted by Fourier transformation. If single crystals of sufficient size cannot be obtained, various other X-ray methods can be applied to obtain less detailed information; such methods include fiber diffraction, powder diffraction and (if the sample is not crystallized) small-angle X-ray scattering (SAXS). If the material under investigation is only available in the form of nanocrystalline powders or suffers from poor crystallinity, the methods of electron crystallography can be applied for determining the atomic structure.

For all above mentioned X-ray diffraction methods, the scattering is elastic; the scattered X-rays have the same wavelength as the incoming X-ray. By contrast, inelastic X-ray scattering methods are useful in studying excitations of the sample, rather than the distribution of its atoms.

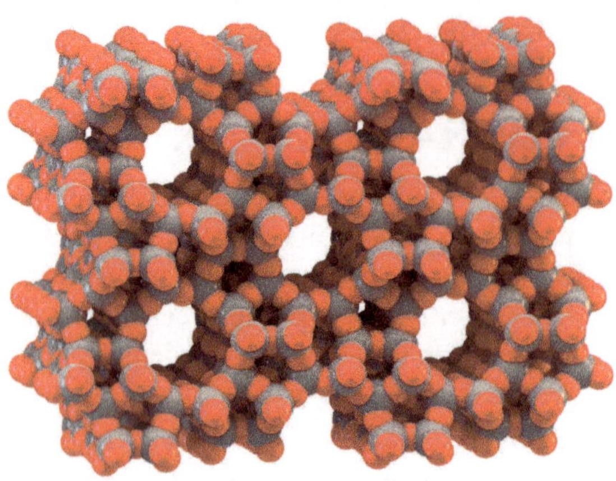

X-ray crystallography provides an atomistic model of zeolite, an aluminosilicate.

Early Scientific History of Crystals and X-Rays

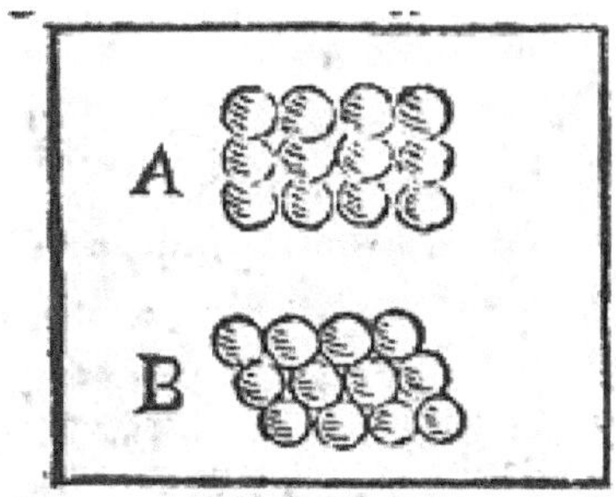

Drawing of square (Figure A, above) and hexagonal (Figure B, below) packing from Kepler's work, *Strena seu de Nive Sexangula.*

Crystals, though long admired for their regularity and symmetry, were not investigated scientifically until the 17th century. Johannes Kepler hypothesized in his work Strena seu de Nive Sexangula (A New Year's Gift of Hexagonal Snow) (1611) that the hexagonal symmetry of snowflake crystals was due to a regular packing of spherical water particles.

As shown by X-ray crystallography, the hexagonal symmetry of snowflakes results from the tetrahedral arrangement of hydrogen bonds about each water molecule. The water molecules are arranged similarly to the silicon atoms in the tridymitepolymorph of SiO_2. The resulting crystal structure has hexagonal symmetry when viewed along a principal axis.

The Danish scientist Nicolas Steno (1669) pioneered experimental investigations of crystal symmetry. Steno showed that the angles between the faces are the same in every exemplar of a particular type of crystal, and René Just Haüy (1784) discovered that every face of a crystal can be described by simple stacking patterns of blocks of the same shape and size. Hence, William Hallowes Miller in 1839 was able to give each face a unique label of three small integers, the Miller indices which remain in use today for identifying crystal faces. Haüy's study led to the correct idea that crystals are a regular three-dimensional array (a Bravais lattice) of atoms and molecules; a single unit cell is repeated indefinitely along three principal directions that are not necessarily perpendicular. In the 19th century, a complete catalog of the possible symmetries of a crystal was worked out by Johan Hessel, Auguste Bravais, Evgraf Fedorov, Arthur Schönflies and (belatedly) William Barlow (1894). From the available data and physical reasoning, Barlow proposed several crystal structures in the 1880s that were validated later by X-ray crystallography; however, the available data were too scarce in the 1880s to accept his models as conclusive.

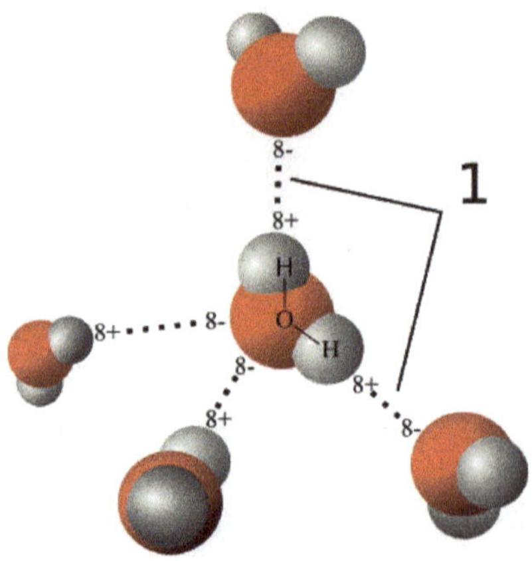

X-ray crystallography shows the arrangement of water molecules in ice, revealing the hydrogen bonds (1) that hold the solid together. Few other methods can determine the structure of matter with such precision (resolution).

Wilhelm Röntgen discovered X-rays in 1895, just as the studies of crystal symmetry were being concluded. Physicists were initially uncertain of the nature of X-rays, but soon suspected (correctly) that they were waves of electromagnetic radiation, in other words, another form of light. At that time, the wave model of light—specifically, the Maxwell theory of electromagnetic radiation—was well accepted among scientists, and experiments by Charles Glover Barkla showed that X-rays exhibited phenomena associated with electromagnetic waves, including transverse polarization and spectral lines akin to those observed in the visible wavelengths. Single-slit experiments in the laboratory of Arnold Sommerfeld suggested that X-rays had a wavelength of about 1 angstrom. However, X-rays are composed of photons, and thus are not only waves of electromagnetic radiation but also exhibit particle-like properties. Albert Einstein introduced the photon concept in 1905, but it was not broadly accepted until 1922, when Arthur Compton confirmed it by the scattering of X-rays from electrons. Therefore, these particle-like properties of X-rays, such as their ionization of gases, caused William Henry Bragg to argue in 1907 that X-rays were not electromagnetic radiation. Nevertheless, Bragg's view was not broadly accepted and the observation of X-ray diffraction by Max von Laue in 1912confirmed for most scientists that X-rays were a form of electromagnetic radiation.

X-Ray Diffraction

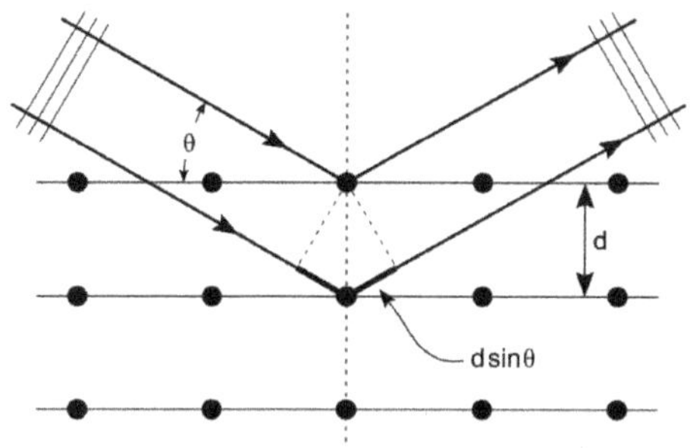

The incoming beam (coming from upper left) causes each scatterer to re-radiate a small portion of its intensity as a spherical wave. If scatterers are arranged symmetrically with a separation d, these spherical waves will be in sync (add constructively) only in directions where their path-length difference 2d sin θ equals an integer multiple of the wavelength λ. In that case, part of the incoming beam is deflected by an angle 2θ, producing a reflection spot in the diffraction pattern.

Crystals are regular arrays of atoms, and X-rays can be considered waves of electromagnetic radiation. Atoms scatter X-ray waves, primarily through the atoms' electrons. Just as an ocean wave striking a lighthouse produces secondary circular waves emanating from the lighthouse, so an X-ray striking an electron produces secondary spherical waves emanating from the electron. This phenomenon is known as elastic scattering, and the electron (or lighthouse) is known as the scatterer. A regular array of scatterers produces a regular array of spherical waves. Although these waves cancel one another out in most directions through destructive interference, they add constructively in a few specific directions, determined by Bragg's law:

$$2d\sin\theta = n\lambda$$

Here d is the spacing between diffracting planes, θ is the incident angle, n is any integer, and λ is the wavelength of the beam. These specific directions appear as spots on the diffraction pattern called reflections. Thus, X-ray diffraction results from an electromagnetic wave (the X-ray) impinging on a regular array of scatterers (the repeating arrangement of atoms within the crystal).

X-rays are used to produce the diffraction pattern because their wavelength λ is typically the same order of magnitude (1–100 angstroms) as the spacing d between planes in the crystal. In principle, any wave impinging on a regular array of scatterers produces diffraction, as predicted first by Francesco Maria Grimaldi in 1665. To produce significant diffraction, the spacing between the scatterers and the wavelength of the impinging wave should be similar in size. For illustration, the diffraction of sunlight through a bird's feather was first reported by James Gregory in the later 17th century. The first artificial diffraction gratings for visible light were constructed by David Rittenhouse in 1787, and Joseph von Fraunhofer in 1821. However, visible light has too long a wavelength (typically, 5500 angstroms) to observe diffraction from crystals. Prior to the first X-ray diffraction experiments, the spacings between lattice planes in a crystal were not known with certainty. The idea that crystals could be used as a diffraction grating for X-rays arose in 1912 in a conversation between Paul Peter Ewald and Max von Laue in the English Garden in Munich. Ewald had proposed a resonator model of crystals for his thesis, but this model could not be validated using visible light, since the wavelength was much larger than the spacing between the resonators. Von Laue realized that electromagnetic radiation of a shorter wavelength was needed to observe such small spacings, and suggested that X-rays might have a wavelength comparable to the unit-cell spacing in crystals. Von Laue worked with two technicians, Walter Friedrich and his assistant Paul Knipping, to shine a beam of X-rays through a copper sulfate crystal and record its diffraction on a photographic plate. After being developed, the plate showed a large number of well-defined spots arranged in a pattern of intersecting circles around the spot produced by the central beam. Von Laue developed a law that connects the scattering angles and the size and orientation of the unit-cell spacings in the crystal, for which he was awarded the Nobel Prize in Physics in 1914.

Scattering

As described in the mathematical derivation below, the X-ray scattering is determined by the density of electrons within the crystal. Since the energy of an X-ray is much greater than that of a valence electron, the scattering may be modeled as Thomson scattering, the interaction of an electromagnetic ray with a free electron. This model is generally adopted to describe the polarization of the scattered radiation.

The intensity of Thomson scattering for one particle with mass m and charge q is:

$$I_o = I_e \left(\frac{q^4}{m^2 c^4}\right) \frac{1+\cos^2 2\theta}{2} = I_e 7.94.10^{-26} \frac{1+\cos^2 2\theta}{2} = I_e f$$

Hence the atomic nuclei, which are much heavier than an electron, contribute negligibly to the scattered X-rays.

Development from 1912 to 1920

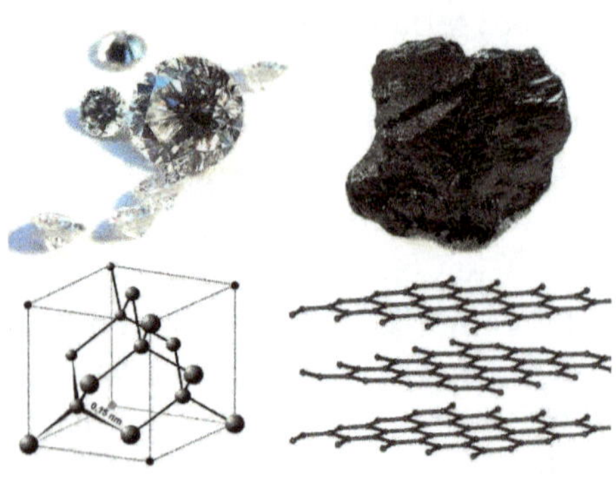

Although diamonds (top left) and graphite (top right) are identical in chemical composition—being both pure carbon—X-ray crystallography revealed the arrangement of their atoms (bottom) accounts for their different properties. In diamond, the carbon atoms are arranged tetrahedrally and held together by single covalent bonds, making it strong in all directions. By contrast, graphite is composed of stacked sheets. Within the sheet, the bonding is covalent and has hexagonal symmetry, but there are no covalent bonds between the sheets, making graphite easy to cleave into flakes.

After Von Laue's pioneering research, the field developed rapidly, most notably by physicists William Lawrence Bragg and his father William Henry Bragg. In 1912–1913, the younger Bragg developed Bragg's law, which connects the observed scattering with reflections from evenly spaced planes within the crystal. The Braggs, father and son, shared the 1915 Nobel Prize in Physics for their work in crystallography. The earliest structures were generally simple and marked by one-dimensional symmetry. However, as computational and experimental methods improved over the next decades, it became feasible to deduce reliable atomic positions for more complicated two- and three-dimensional arrangements of atoms in the unit-cell.

The potential of X-ray crystallography for determining the structure of molecules and minerals—then only known vaguely from chemical and hydrodynamic experiments—was realized immediately. The earliest structures were simple inorganic crystals and minerals, but even these revealed fundamental laws of physics and chemistry. The first atomic-resolution structure to be "solved" (i.e., determined) in 1914 was that of table salt. The distribution of electrons in the table-salt structure showed that crystals are not necessarily composed of covalently bonded molecules, and proved the existence of ionic compounds. The structure of diamond was solved in the same year, proving the tetrahedral arrangement of its chemical bonds and showing that the length of C–C single bond was 1.52 angstroms. Other early structures included copper, calcium fluoride (CaF_2, also known as fluorite), calcite ($CaCO_3$) and pyrite (FeS_2) in 1914; spinel ($MgAl_2O_4$) in 1915; the rutile and anatase forms of titanium dioxide (TiO_2) in 1916; pyrochroite $Mn(OH)_2$ and, by extension, brucite $Mg(OH)_2$ in 1919;.Also in 1919 sodium nitrate ($NaNO_3$) and caesium dichloroiodide ($CsICl_2$) were determined by Ralph Walter Graystone Wyckoff, and the wurtzite (hexagonal ZnS) structure became known in 1920.

The structure of graphite was solved in 1916 by the related method of powder diffraction, which was developed by Peter Debye and Paul Scherrer and, independently, by Albert Hull in 1917. The structure of graphite was determined from single-crystal diffraction in 1924 by two groups independently. Hull also used the powder method to determine the structures of various metals, such as iron and magnesium.

Cultural and Aesthetic Importance of X-Ray Crystallography

In what has been called his scientific autobiography, The Development of X-ray Analysis, Sir William Lawrence Bragg mentioned that he believed the field of crystallography was particularly welcoming to women because the techno-aesthetics of the molecular structures resembled textiles and household objects. Bragg was known to compare crystal formation to "curtains, wallpapers, mosaics, and roses." In 1951, the Festival Pattern Group at the Festival of Britain hosted a collaborative group of textile manufacturers and experienced crystallographers to design lace and prints based on the X-ray crystallography of insulin, china clay, and hemoglobin. One of the leading scientists of the project was Dr. Helen Megaw (1907–2002), the Assistant Director of Research at the Cavendish Laboratory in Cambridge at the time. Megaw is credited as one of the central figures who took inspiration from crystal diagrams and saw their potential in design. In 2008, the Wellcome Collection in London curated an exhibition on the Festival Pattern Group called "From Atom to Patterns."

Contributions to Chemistry and Material Science

X-ray crystallography has led to a better understanding of chemical bonds and non-covalent interactions. The initial studies revealed the typical radii of atoms, and confirmed many theoretical models of chemical bonding, such as the tetrahedral bonding of carbon in the diamond structure, the octahedral bonding of metals observed in ammonium hexachloroplatinate (IV), and the resonance observed in the planar carbonate group and in aromatic molecules. Kathleen Lonsdale's 1928 structure of hexamethylbenzene established the hexagonal symmetry of benzene and showed a clear difference in bond length between the aliphatic C–C bonds and aromatic C–C bonds; this finding led to the idea of resonance between chemical bonds, which had profound consequences for the development of chemistry. Her conclusions were anticipated by William Henry Bragg, who published models of naphthalene and anthracene in 1921 based on other molecules, an early form of molecular replacement.

Also in the 1920s, Victor Moritz Goldschmidt and later Linus Pauling developed rules for eliminating chemically unlikely structures and for determining the relative sizes of atoms. These rules led to the structure of brookite (1928) and an understanding of the relative stability of the rutile, brookite and anatase forms of titanium dioxide.

The distance between two bonded atoms is a sensitive measure of the bond strength and its bond order; thus, X-ray crystallographic studies have led to the discovery of even more exotic types of bonding in inorganic chemistry, such as metal-metal double bonds, metal-metal quadruple bonds, and three-center, two-electron bonds. X-ray crystallography—or, strictly speaking, an inelastic Compton scattering experiment—has also provided evidence for the partly covalent character of hydrogen bonds. In the field of organometallic chemistry, the X-ray structure of ferrocene initiated scientific studies of sandwich compounds, while that of Zeise's salt stimulated research into "back bonding" and metal-pi complexes. Finally, X-ray crystallography had a pioneering role in the development of supramolecular chemistry, particularly in clarifying the structures of the crown ethers and the principles of host-guest chemistry.

In material sciences, many complicated inorganic and organometallic systems have been analyzed using single-crystal methods, such as fullerenes, metalloporphyrins, and other complicated compounds. Single-crystal diffraction is also used in the pharmaceutical industry, due to recent problems with polymorphs. The major factors affecting the quality of single-crystal structures are the crystal's size and regularity; recrystallization is a commonly used technique to improve these factors in small-molecule crystals. The Cambridge Structural Database contains over 800,000 structures as of September 2016; over 99% of these structures were determined by X-ray diffraction.

Mineralogy and Metallurgy

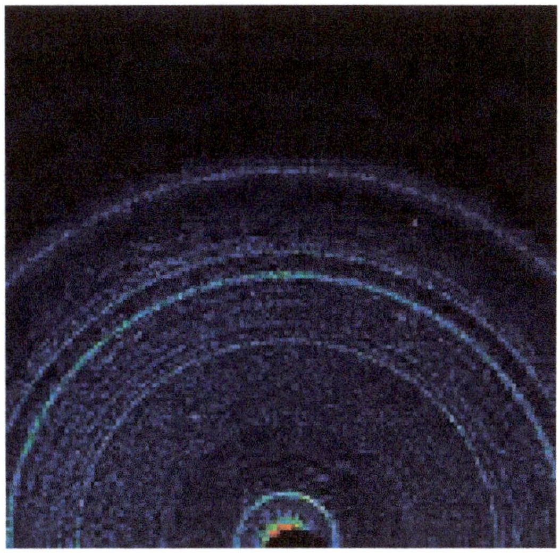

First X-ray diffraction view of Martian soil – CheMin analysis reveals feldspar, pyroxenes, olivine and more (Curiosity rover at "Rocknest", October 17, 2012).

Since the 1920s, X-ray diffraction has been the principal method for determining the arrangement of atoms in minerals and metals. The application of X-ray crystallography to mineralogy began with the structure of garnet, which was determined in 1924 by Menzer. A systematic X-ray crystallographic study of the silicates was undertaken in the 1920s. This study showed that, as the Si/O ratio is altered, the silicate crystals exhibit significant changes in their atomic arrangements. Machatschki extended these insights to minerals in which aluminium substitutes for the silicon atoms of the silicates. The first application of X-ray crystallography to metallurgy likewise occurred in the mid-1920s. Most notably, Linus Pauling's structure of the alloy Mg_2Sn led to his theory of the stability and structure of complex ionic crystals.

On October 17, 2012, the Curiosity rover on the planet Mars at "Rocknest" performed the first X-ray diffraction analysis of Martian soil. The results from the rover's CheMin analyzer revealed the presence of several minerals, including feldspar, pyroxenes and olivine, and suggested that the Martian soil in the sample was similar to the "weathered basaltic soils" of Hawaiian volcanoes.

Early Organic and Small Biological Molecules

The first structure of an organic compound, hexamethylenetetramine, was solved in 1923. This was followed by several studies of long-chain fatty acids, which are an important component of biological membranes. In the 1930s, the structures of much larger molecules with two-dimensional complexity began to be solved. A significant advance was the structure of phthalocyanine, a large planar molecule that is closely related to porphyrin molecules important in biology, such as heme, corrin and chlorophyll.

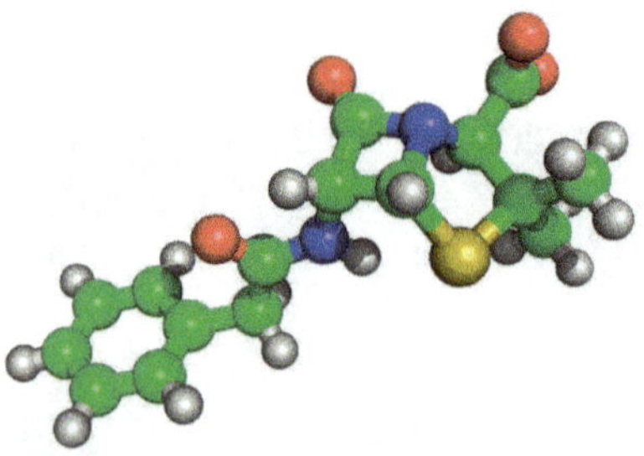

The three-dimensional structure of penicillin, solved by Dorothy Crowfoot Hodgkin in 1945. The green, white, red, yellow and blue spheres represent atoms of carbon, hydrogen, oxygen, sulfur and nitrogen, respectively.

X-ray crystallography of biological molecules took off with Dorothy Crowfoot Hodgkin, who solved the structures of cholesterol (1937), penicillin (1946) and vitamin B12 (1956), for which she was awarded the Nobel Prize in Chemistry in 1964. In 1969, she succeeded in solving the structure of insulin, on which she worked for over thirty years.

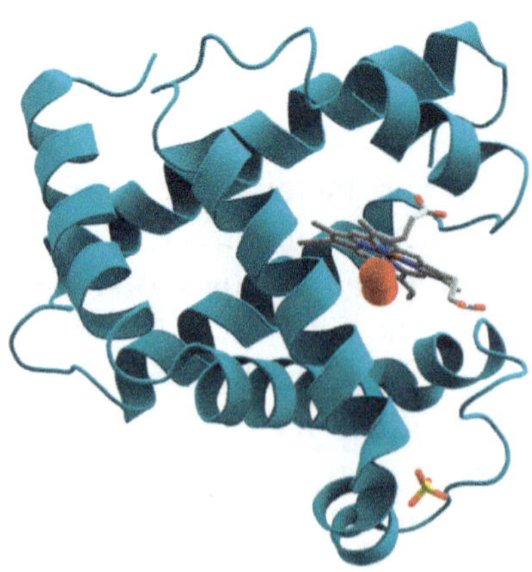

Ribbon diagram of the structure of myoglobin, showing colored alpha helices. Such proteins are long, linear molecules with thousands of atoms; yet the relative position of each atom has been determined with sub-atomic resolution by X-ray crystallography. Since it is difficult to visualize all the atoms at once, the ribbon shows the rough path of the protein polymer from its N-terminus (blue) to its C-terminus (red).

Biological Macromolecular Crystallography

Crystal structures of proteins (which are irregular and hundreds of times larger than cholesterol) began to be solved in the late 1950s, beginning with the structure of sperm whale myoglobin by Sir John Cowdery Kendrew, for which he shared the Nobel Prize in Chemistry with Max Perutz in 1962. Since that success, over 86817 X-ray crystal structures of proteins, nucleic acids and other biological molecules have been determined. For comparison, the nearest competing method in terms of structures analyzed is nuclear magnetic resonance (NMR) spectroscopy, which has resolved 9561 chemical structures. Moreover, crystallography can solve structures of arbitrarily large molecules, whereas solution-state NMR is restricted to relatively small ones (less than 70 kDa). X-ray crystallography is now used routinely by scientists to determine how a pharmaceutical drug interacts with its protein target and what changes might improve it. However, intrinsic membrane proteins remain challenging to crystallize because they require detergents or other means to solubilize them in isolation, and such detergents often interfere with crystallization. Such membrane proteins are a large component of the genome, and include many proteins of great physiological importance, such as ion channels and receptors. Helium cryogenics are used to prevent radiation damage in protein crystals.

On the other end of the size scale, relative small molecules are able to lure the resolving power of X-ray crystallography. The structure assigned in 1991 to the antibiotic isolated from a marina organism, diazonamide A – $C_{40}H_{34}Cl_2N_6O_6$ with M = 765.65 – proved to be incorrect by the classical proof of structure: a synthetic sample was not identical to the natural product. The mistake was possible because of the inability of X-ray crystallography to distinguish between the correct -OH / >NH and the interchanged $-NH_2$ / -O- groups in the incorrect structure.

Relationship to Other Scattering Techniques

Elastic Vs. Inelastic Scattering

X-ray crystallography is a form of elastic scattering; the outgoing X-rays have the same energy, and thus same wavelength, as the incoming X-rays, only with altered direction. By contrast, inelastic scattering occurs when energy is transferred from the incoming X-ray to the crystal, e.g., by exciting an inner-shell electron to a higher energy level. Such inelastic scattering reduces the energy (or increases the wavelength) of the outgoing beam. Inelastic scattering is useful for probing such excitations of matter, but not in determining the distribution of scatterers within the matter, which is the goal of X-ray crystallography.

X-rays range in wavelength from 10 to 0.01 nanometers; a typical wavelength used for crystallography is 1 Å (0.1 nm), which is on the scale of covalent chemical bonds and the radius of a single atom. Longer-wavelength photons (such as ultraviolet radiation) would not have sufficient resolution to determine the atomic positions. At the other extreme, shorter-wavelength photons such as gamma rays are difficult to produce in large numbers, difficult to focus, and interact too strongly with matter, producing particle-antiparticle pairs. Therefore, X-rays are the "sweetspot" for wavelength when determining atomic-resolution structures from the scattering of electromagnetic radiation.

Other X-Ray Techniques

Other forms of elastic X-ray scattering include powder diffraction, Small-Angle X-ray Scattering (SAXS) and several types of X-ray fiber diffraction, which was used by Rosalind Franklin in determining the double-helix structure of DNA. In general, single-crystal X-ray diffraction offers more structural information than these other techniques;

however, it requires a sufficiently large and regular crystal, which is not always available.

These scattering methods generally use monochromatic X-rays, which are restricted to a single wavelength with minor deviations. A broad spectrum of X-rays (that is, a blend of X-rays with different wavelengths) can also be used to carry out X-ray diffraction, a technique known as the Laue method. This is the method used in the original discovery of X-ray diffraction. Laue scattering provides much structural information with only a short exposure to the X-ray beam, and is therefore used in structural studies of very rapid events (Time resolved crystallography). However, it is not as well-suited as monochromatic scattering for determining the full atomic structure of a crystal and therefore works better with crystals with relatively simple atomic arrangements.

The Laue back reflection mode records X-rays scattered backwards from a broad spectrum source. This is useful if the sample is too thick for X-rays to transmit through it. The diffracting planes in the crystal are determined by knowing that the normal to the diffracting plane bisects the angle between the incident beam and the diffracted beam. A Greninger chart can be usedto interpret the back reflection Laue photograph.

Electron and Neutron Diffraction

Other particles, such as electrons and neutrons, may be used to produce a diffraction pattern. Although electron, neutron, and X-ray scattering are based on different physical processes, the resulting diffraction patterns are analyzed using the same coherent diffraction imaging techniques.

As derived below, the electron density within the crystal and the diffraction patterns are related by a simple mathematical method, the Fourier transform, which allows the density to be calculated relatively easily from the patterns. However, this works only if the scattering is weak, i.e., if the scattered beams are much less intense than the incoming beam. Weakly scattered beams pass through the remainder of the crystal without undergoing a second scattering event. Such re-scattered waves are called "secondary scattering" and hinder the analysis. Any sufficiently thick crystal will produce secondary scattering, but since X-rays interact relatively weakly with the electrons, this is generally not a significant concern. By contrast, electron beams may produce strong secondary scattering even for relatively thin crystals (>100 nm). Since this thickness corresponds to the diameter of many viruses, a promising direction is the electron diffraction of isolated macromolecular assemblies, such as viral capsids and molecular machines, which may be carried out with a cryo-electron microscope. Moreover, the strong interaction of electrons with matter (about 1000 times stronger than for X-rays) allows determination of the atomic structure of extremely small volumes. The field of applications for electron crystallography ranges from bio molecules like membrane proteins over organic thin films to the complex structures of (nanocrystalline) intermetallic compounds and zeolites.

Neutron diffraction is an excellent method for structure determination, although it has been difficult to obtain intense, monochromatic beams of neutrons in sufficient quantities. Traditionally, nuclear reactors have been used, although sources producing neutrons by spallation are becoming increasingly available. Being uncharged, neutrons scatter much more readily from the atomic nuclei rather than from the electrons. Therefore, neutron scattering is very useful for observing the positions of light atoms with few electrons, especially hydrogen, which is essentially invisible in the X-ray diffraction. Neutron scattering also has the remarkable property that the solvent can be made invisible by adjusting the ratio of normal water, H_2O, and heavy water, D_2O.

Methods

Overview of Single-Crystal X-Ray Diffraction

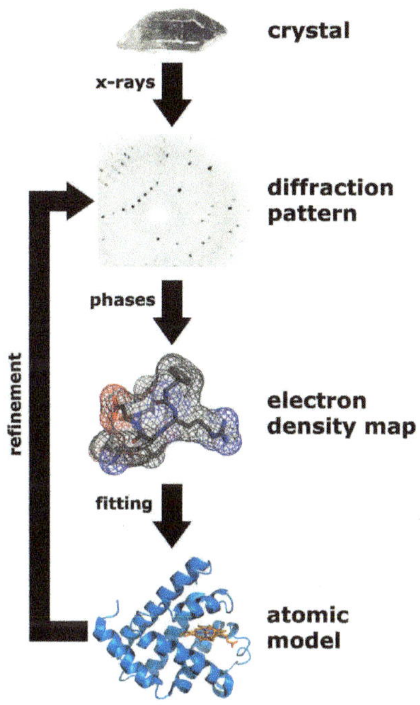

Workflow for solving the structure of a molecule by X-ray crystallography.

The oldest and most precise method of X-ray crystallography is single-crystal X-ray diffraction, in which a beam of X-rays strikes a single crystal, producing scattered beams. When they land on a piece of film or other detector, these beams make a diffraction pattern of spots; the strengths and angles of these beams are recorded as the crystal is gradually rotated. Each spot is called a reflection, since it corresponds to the reflection of the X-rays from one set of evenly spaced planes within the crystal. For single crystals of sufficient purity and regularity, X-ray diffraction data can determine the mean chemical bond lengths and angles to within a few thousandths of an angstrom and to within a few tenths of a degree, respectively. The atoms in a crystal are not static, but oscillate about their mean positions, usually by less than a few tenths of an angstrom. X-ray crystallography allows measuring the size of these oscillations.

Procedure

The technique of single-crystal X-ray crystallography has three basic steps. The first—and often most difficult—step is to obtain an adequate crystal of the material under study. The crystal should be sufficiently large (typically larger than 0.1 mm in all dimensions), pure in composition and regular in structure, with no significant internal imperfections such as cracks or twinning.

In the second step, the crystal is placed in an intense beam of X-rays, usually of a single wavelength (monochromatic X-rays), producing the regular pattern of reflections. As the crystal is gradually rotated, previous reflections disappear and new ones appear; the intensity of every spot is recorded at every orientation of the crystal. Multiple data sets may have to be collected, with each set covering slightly more than half a full rotation of the crystal and typically containing tens of thousands of reflections.

In the third step, these data are combined computationally with complementary chemical information to produce and refine a model of the arrangement of atoms within the crystal. The final, refined model of the atomic arrangement—now called a crystal structure—is usually stored in a public database.

Limitations

As the crystal's repeating unit, its unit cell, becomes larger and more complex, the atomic-level picture provided by X-ray crystallography becomes less well-resolved (more "fuzzy") for a given number of observed reflections. Two limiting cases of X-ray crystallography—"small-molecule" (which includes continuous inorganic solids) and "macromolecular" crystallography—are often discerned. Small-molecule crystallography typically involves crystals with fewer than 100 atoms in their asymmetric unit; such crystal structures are usually so well resolved that the atoms can be discerned as isolated "blobs" of electron density. By contrast, macromolecular crystallography often involves tens of thousands of atoms in the unit cell. Such crystal structures are generally less well-resolved (more "smeared out"); the atoms and chemical bonds appear as tubes of electron density, rather than as isolated atoms. In general, small molecules are also easier to crystallize than macromolecules; however, X-ray crystallography has proven possible even for viruses with hundreds of thousands of atoms. Though normally x-ray crystallography can only be performed if the sample is in crystal form, new research has been done into sampling non-crystalline forms of samples.

Crystallization

A protein crystal seen under a microscope. Crystals used in X-ray crystallography may be smaller than a millimeter across.

Although crystallography can be used to characterize the disorder in an impure or irregular crystal, crystallography generally requires a pure crystal of high regularity to solve the structure of a complicated arrangement of atoms. Pure, regular crystals can sometimes be obtained from natural or synthetic materials, such as samples of metals, minerals or other macroscopic materials. The regularity of such crystals can sometimes be improved with macromolecular crystal annealing and other methods. However, in many cases, obtaining a diffraction-quality crystal is the chief barrier to solving its atomic-resolution structure.

Small-molecule and macromolecular crystallography differ in the range of possible techniques used to produce diffraction-quality crystals. Small molecules generally have few degrees of conformational freedom, and may be crystallized by a wide range of methods, such as chemical vapor deposition and recrystallization. By contrast, macromolecules generally have many degrees of freedom and their crystallization must be carried out so as to maintain a stable structure. For example, proteins and larger RNA molecules cannot be crystallized if their tertiary structure has been unfolded; therefore, the range of crystallization conditions is restricted to solution conditions in which such molecules remain folded.

Protein crystals are almost always grown in solution. The most common approach is to lower the solubility of its component molecules very gradually; if this is done too quickly, the molecules will precipitate from solution, forming a useless dust or amorphous gel on the bottom of the container. Crystal growth in solution is characterized by two steps: nucleation of a microscopic crystallite (possibly having only 100 molecules), followed by growth of that crystallite, ideally to a diffraction-quality crystal. The solution conditions that favor the first step (nucleation) are not always the same conditions that favor the second step (subsequent growth). The crystallographer's goal is to identify solution conditions that favor the development of a single, large crystal, since larger crystals offer improved resolution of the molecule. Consequently, the solution conditions should disfavor the first step (nucleation) but favor the second (growth), so that only one large crystal forms per droplet. If nucleation is favored too much, a shower of small crystallites will form in the droplet, rather than one large crystal; if favored too little, no crystal will form whatsoever. Other approaches involves, crystallizing proteins under oil, where aqueous protein solutions are dispensed under liquid oil, and water evaporates through the layer of oil. Different oils have different evaporation permeabilities, therefore yielding changes in concentration rates from different percipient/protein mixture. The technique relies on bringing the protein directly into the nucleation zone by mixing protein with the appropriate amount of percipientto prevent the diffusion of water out of the drop.

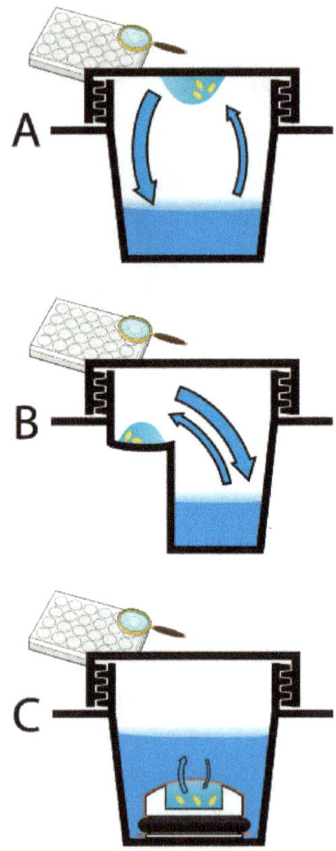

Three methods of preparing crystals, A: Hanging drop. B: Sitting drop. C: Microdialysis

It is extremely difficult to predict good conditions for nucleation or growth of well-ordered crystals. In practice, favorable conditions are identified by screening; a very large batch of the molecules is prepared, and a wide variety of crystallization solutions are tested.Hundreds, even thousands, of solution conditions are generally tried before finding the successful one. The various conditions can use one or more physical mechanisms to lower the solubility of the molecule; for example, some may change the pH, some contain salts of the Hofmeister series or chemicals that lower the dielectric constant of the solution, and still others contain large polymers such as polyethylene glycol that drive the molecule out of solution by entropic effects. It is also common to try several temperatures for encouraging crystallization, or to gradually lower the temperature so that the solution becomes supersaturated. These methods require large amounts of the target molecule, as they use high concentration of the molecule(s) to be crystallized. Due to the difficulty in obtaining such large quantities (milligrams) of crystallization-grade protein, robots have been developed that are capable of accurately dispensing crystallization trial drops that are in the order of 100 nanoliters in volume. This means that 10-fold less protein is used per experiment when compared to crystallization trials set up by hand (in the order of 1 microliter).

Several factors are known to inhibit or mar crystallization. The growing crystals are generally held at a constant temperature and protected from shocks or vibrations that might disturb their crystallization. Impurities in the molecules or in the crystallization solutions are often inimical to crystallization. Conformational flexibility in the molecule also tends to make crystallization less likely, due to entropy. Ironically, molecules that tend to self-assemble into regular helices are often unwilling to assemble into crystals. Crystals can be marred by twinning, which can occur when a unit cell can pack equally favorably in multiple orientations; although recent advances in computational methods may allow solving the structure of some twinned crystals. Having failed to crystallize a target molecule, a crystallographer may try again with a slightly modified version of the molecule; even small changes in molecular properties can lead to large differences in crystallization behavior.

Data Collection

Mounting the Crystal

Screen capture from animation showing the five motions possible with a four-circle kappa goniometer. The rotations about each of the four angles φ, κ, ω and 2θ leave the crystal within the X-ray beam, but change the crystal orientation. The detector (red box) can be slid closer or further away from the crystal, allowing higher resolution data to be taken (if closer) or better discernment of the Bragg peaks (if further away).

The crystal is mounted for measurements so that it may be held in the X-ray beam and rotated. There are several methods of mounting. In the past, crystals were loaded into glass capillaries with the crystallization solution (the mother liquor). Nowadays, crystals of small molecules are typically attached with oil or glue to a glass fiber or a loop, which is made of nylon or plastic and attached to a solid rod. Protein crystals are scooped up by a loop, then flash-frozen with liquid nitrogen. This freezing reduces the radiation damage of the X-rays, as well as the noise in the Bragg peaks due to thermal motion (the Debye-Waller effect). However, untreated protein crystals often crack if flash-frozen; therefore, they are generally pre-soaked in a cryoprotectant solution before freezing. Unfortunately, this pre-soak may itself cause the crystal to crack, ruining it for crystallography. Generally, successful cryo-conditions are identified by trial and error.

The capillary or loop is mounted on a goniometer, which allows it to be positioned accurately within the X-ray beam and rotated. Since both the crystal and the beam are often very small, the crystal must be centered within the beam to within ~25 micrometers accuracy, which is aided by a camera focused on the crystal. The most common type of goniometer is the "kappa goniometer", which offers three angles of rotation: the ω angle, which rotates about an axis perpendicular to the beam; the κ angle, about an axis at ~50° to the ω axis; and, finally, the φ angle about the loop/capillary axis. When the κ angle is zero, the ω and φ axes are aligned. The κ rotation allows for convenient mounting of the crystal, since the arm in which the crystal is mounted may be swung out towards the crystallographer. The oscillations carried out during data collection (mentioned below) involve the ω axis only. An older type of goniometer is the four-circle goniometer, and its relatives such as the six-circle goniometer.

X-Ray Sources

Rotating Anode

Small scale can be done on a local X-ray tube source, typically coupled with an image plate detector. These have the advantage of being (relatively) inexpensive and easy to maintain, and allow for quick screening and collection of samples. However, the wavelength light produced is limited by anode material, typically copper. Further, intensity is limited by the power applied and cooling capacity available to avoid melting the anode. In such systems, electrons are boiled off of a cathode and accelerated through a strong electric potential of ~50 kV; having reached a high speed, the electrons collide with a metal plate, emitting bremsstrahlung and some strong spectral lines corresponding to the excitation of inner-shell electrons of the metal. The most common metal used is copper, which can be kept cool easily, due to its high thermal conductivity, and which produces strong K_α and K_β lines. The K_β line is sometimes suppressed with a thin (~10 μm) nickel foil. The simplest and cheapest variety of sealed X-ray tube has a stationary anode (the Crookes tube) and run with ~2 kW of electron beam power. The more expensive variety has a rotating-anode type source that run with ~14 kW of e-beam power. X-rays are generally filtered (by use of X-Ray Filters) to a single wavelength (made monochromatic) and collimated to a single direction before they are allowed to strike the crystal. The filtering not only simplifies the data analysis, but also removes radiation that degrades the crystal without contributing useful information. Collimation is done either with a collimator (basically, a long tube) or with a clever arrangement of gently curved mirrors. Mirror systems are preferred for small crystals (under 0.3 mm) or with large unit cells (over 150 Å).

Synchrotron Radiation

Synchrotron radiation are some of the brightest lights on earth. It is the single most powerful tool available to X-ray crystallographers. It is made of X-ray beams generated in large machines called synchrotrons. These machines accelerate electrically charged particles, often electrons, to nearly the speed of light and confine them in a (roughly) circular loop using magnetic fields.

Synchrotrons are generally national facilities, each with several dedicated beamlines where data is collected without interruption. Synchrotrons were originally designed for use by high-energy physicists studying subatomic particles and cosmic phenomena. The largest component of each synchrotron is its electron storage ring. This ring is actually not a perfect circle, but a many-sided polygon. At each corner of the polygon, or sector, precisely aligned magnets bend the electron stream. As the electrons' path is bent, they emit bursts of energy in the form of X-rays.

Using synchrotron radiation frequently has specific requirements for X-ray crystallography. The intense ionizing radiation can cause radiation damage to samples, particularly macromolecular crystals. Cryo crystallography protects the sample from radiation damage, by freezing the crystal at liquid nitrogen temperatures (~100 K). However, synchrotron radiation frequently has the advantage of user selectable wavelengths, allowing for anomalous scattering experiments which maximizes anomalous signal. This is critical in experiments such as SAD and MAD.

Free Electron Laser

Recently, free electron lasers have been developed for use in X-ray crystallography. These are the brightest X-ray sources currently available; with the X-rays coming in femtosecond bursts. The intensity of the source is such that atomic resolution diffraction patterns can be resolved for crystals otherwise too small for collection. However, the intense light source also destroys the sample, requiring multiple crystals to be shot. As each crystal is randomly oriented in the beam, hundreds of thousands of individual diffraction images must be collected in order to get a complete data-set. This method, serial femtosecond crystallography, has been used in solving the structure of a number of protein crystal structures, sometimes noting differences with equivalent structures collected from synchrotron sources.

Recording the Reflections

When a crystal is mounted and exposed to an intense beam of X-rays, it scatters the X-rays into a pattern of spots or reflections that can be observed on a screen behind the crystal. A similar pattern may be seen by shining a laser pointer at a compact disc. The relative intensities of these spots provide the information to determine the arrangement of molecules within the crystal in atomic detail. The intensities of these reflections may be recorded with photographic film, an area detector or with a charge-coupled device (CCD) image sensor. The peaks at small angles correspond to low-resolution data, whereas those at high angles represent high-resolution data; thus, an upper limit on the eventual resolution of the structure can be determined from the first few images. Some measures of diffraction quality can be determined at this point, such as the mosaicity of the crystal and its overall disorder, as observed in the peak widths. Some pathologies of the crystal that would render it unfit for solving the structure can also be diagnosed quickly at this point.

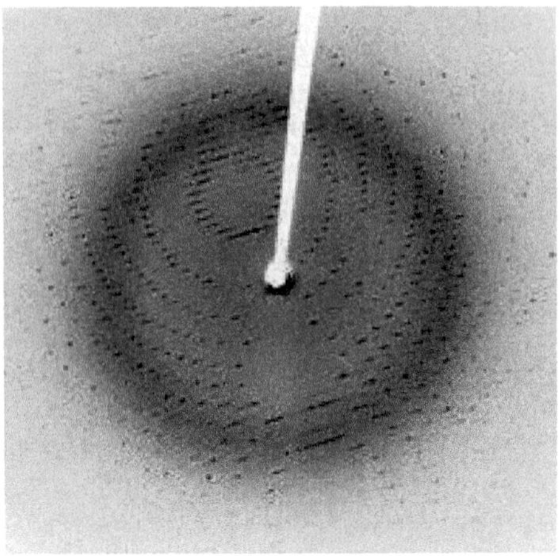

An X-ray diffraction pattern of a crystallized enzyme. The pattern of spots (reflections) and the relative strength of each spot (intensities) can be used to determine the structure of the enzyme.

One image of spots is insufficient to reconstruct the whole crystal; it represents only a small slice of the full Fourier transform. To collect all the necessary information, the crystal must be rotated step-by-step through 180°, with an image recorded at every step; actually, slightly more than 180° is required to cover reciprocal space, due to the curvature of the Ewald sphere. However, if the crystal has a higher symmetry, a smaller angular range such as 90° or 45° may be recorded. The rotation axis should be changed at least once, to avoid developing a "blind spot" in reciprocal space close to the rotation axis. It is customary to rock the crystal slightly (by 0.5–2°) to catch a broader region of reciprocal space.

Multiple data sets may be necessary for certain phasing methods. For example, MAD phasing requires that the scattering be recorded at least three (and usually four, for redundancy) wavelengths of the incoming X-ray radiation. A single crystal may degrade too much during the collection of one data set, owing to radiation damage; in such cases, data sets on multiple crystals must be taken.

Data Analysis

Crystal Symmetry, Unit Cell, and Image Scaling

The recorded series of two-dimensional diffraction patterns, each corresponding to a different crystal orientation, is converted into a three-dimensional model of the electron density; the conversion uses the mathematical technique of Fourier transforms, which is explained below. Each spot corresponds to a different type of variation in the electron density; the crystallographer must determine which variation corresponds to which spot (indexing), the relative strengths of the spots in different images (merging and scaling) and how the variations should be combined to yield the total electron density (phasing).

Data processing begins with indexing the reflections. This means identifying the dimensions of the unit cell and which image peak corresponds to which position in reciprocal space. A byproduct of indexing is to determine the symmetry of the crystal, i.e., its space group. Some space groups can be eliminated from the beginning. For example, reflection symmetries cannot be observed in chiral molecules; thus, only 65 space groups of 230 possible are allowed for protein molecules which are almost always chiral. Indexing is generally accomplished using an autoindexing routine. Having assigned symmetry, the data is then integrated. This converts the hundreds of images containing the thousands of reflections into a single file, consisting of (at the very least) records of the Miller index of each reflection, and an intensity for each reflection (at this state the file often also includes error estimates and measures of partiality (what part of a given reflection was recorded on that image)).

A full data set may consist of hundreds of separate images taken at different orientations of the crystal. The first step is to merge and scale these various images, that is, to identify which peaks appear in two or more images (merging) and to scale the relative images so that they have a consistent intensity scale. Optimizing the intensity scale is critical because the relative intensity of the peaks is the key information from which the structure is determined. The repetitive technique of crystallographic data collection and the often high symmetry of crystalline materials cause the diffractometer to record many symmetry-equivalent reflections multiple times. This allows calculating the symmetry-related R-factor, a reliability index based upon how similar are the measured intensities of symmetry-equivalent reflections, thus assessing the quality of the data.

Initial Phasing

The data collected from a diffraction experiment is a reciprocal space representation of the crystal lattice. The position of each diffraction 'spot' is governed by the size and shape of the unit cell, and the inherent symmetry within the crystal. The intensity of each diffraction 'spot' is recorded, and this intensity is proportional to the square of the structure factor amplitude. The structure factor is a complex number containing information relating to both the amplitude and phase of a wave. In order to obtain an interpretable electron density map, both amplitude and phase must be known (an electron density map allows a crystallographer to build a starting model of the molecule). The phase cannot be directly recorded during a diffraction experiment: this is known as the phase problem. Initial phase estimates can be obtained in a variety of ways:

- Ab initio **phasing** or direct methods – This is usually the method of choice for small molecules (<1000 non-hydrogen atoms), and has been used successfully to solve the phase problems for small proteins. If the resolution of the data is better than 1.4 Å (140 pm), direct methods can be used to obtain phase information, by exploiting known phase relationships between certain groups of reflections.
- Molecular replacement – if a related structure is known, it can be used as a

search model in molecular replacement to determine the orientation and position of the molecules within the unit cell. The phases obtained this way can be used to generate electron density maps.

- Anomalous X-ray scattering (MAD or SAD phasing) – the X-ray wavelength may be scanned past an absorption edge of an atom, which changes the scattering in a known way. By recording full sets of reflections at three different wavelengths (far below, far above and in the middle of the absorption edge) one can solve for the substructure of the anomalously diffracting atoms and hence the structure of the whole molecule. The most popular method of incorporating anomalous scattering atoms into proteins is to express the protein in a methionine auxotroph (a host incapable of synthesizing methionine) in a media rich in seleno-methionine, which contains selenium atoms. A MAD experiment can then be conducted around the absorption edge, which should then yield the position of any methionine residues within the protein, providing initial phases.
- Heavy atom methods (multiple isomorphous replacement) – If electron-dense metal atoms can be introduced into the crystal, direct methods or Patterson-space methods can be used to determine their location and to obtain initial phases. Such heavy atoms can be introduced either by soaking the crystal in a heavy atom-containing solution, or by co-crystallization (growing the crystals in the presence of a heavy atom). As in MAD phasing, the changes in the scattering amplitudes can be interpreted to yield the phases. Although this is the original method by which protein crystal structures were solved, it has largely been superseded by MAD phasing with selenomethionine.

Model Building and Phase Refinement

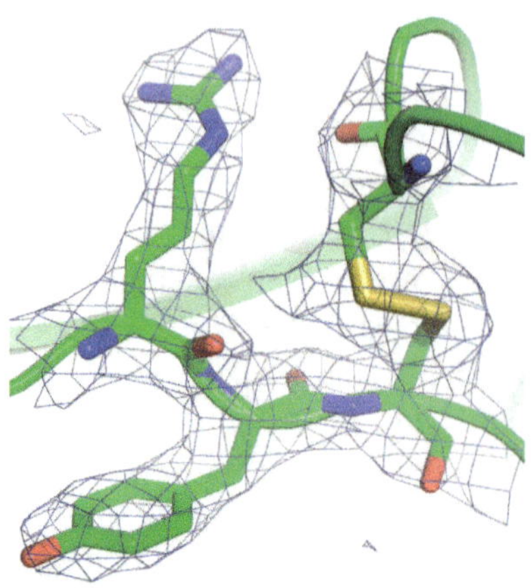

A protein crystal structure at 2.7 Å resolution. The mesh encloses the region in which the electron density exceeds a given threshold. The straight segments represent chemical bonds between the non-hydrogen atoms of an arginine (upper left), a tyrosine (lower left), a disulfide bond (upper right, in yellow), and some peptide groups (running left-right in the middle). The two curved green tubes represent spline fits to the polypeptide backbone.

Having obtained initial phases, an initial model can be built. This model can be used to refine the phases, leading to an improved model, and so on. Given a model of some atomic positions, these positions and their respective Debye-Waller factors (or **B**-factors, accounting for the thermal motion of the atom) can be refined to fit the observed diffraction data, ideally yielding a better set of phases. A new model can then be fit to the new electron density map and a further round of refinement is carried out. This continues until the correlation between the diffraction data and the model is maximized. The agreement is measured by an R-factor defined as

$$R = \frac{\sum_{\text{all reflections}} |F_o - F_c|}{\sum_{\text{all reflections}} |F_o|}$$

where F is the structure factor. A similar quality criterion is R_{free}, which is calculated from a subset (~10%) of reflections that were not included in the structure refinement. Both R factors depend on the resolution of the data. As a rule of thumb, R_{free} should be approximately the resolution in angstroms divided by 10; thus, a data-set with 2 Å resolution should yield a final $R_{free} \sim 0.2$. Chemical bonding features such as stereochemistry, hydrogen bonding and distribution of bond lengths and angles are complementary measures of the model quality. Phase bias is a serious problem in such iterative model building. Omit maps are a common technique used to check for this.

It may not be possible to observe every atom in the asymmetric unit. In many cases, disorder smears the electron density map. Weakly scattering atoms such as hydrogen are routinely invisible. It is also possible for a single atom to appear multiple times in an electron density map, e.g., if a protein sidechain has multiple (<4) allowed conformations. In still other cases, the crystallographer may detect that the covalent structure deduced for the molecule was incorrect, or changed. For example, proteins may be cleaved or undergo post-translational modifications that were not detected prior to the crystallization.

Disorder

A common challenge in refinement of crystal structures results from crystallographic disorder. Disorder can take many forms but in general involves the coexistence of two or more species or conformations. Failure to recognize disorder results in flawed interpretation. Pitfalls from improper modeling of disorder are illustrated by the discounted hypothesis of bond stretch isomerism. Disorder is modelled with respect to the relative population of the components, often only two, and their identity. In structures of large molecules and ions, solvent and counterions are often disordered.

Deposition of the Structure

Once the model of a molecule's structure has been finalized, it is often deposited in a crystallographic database such as the Cambridge Structural Database (for small molecules), the Inorganic Crystal Structure Database (ICSD) (for inorganic compounds) or the Protein Data Bank (for protein structures). Many structures obtained in private commercial ventures to crystallize medicinally relevant proteins are not deposited in public crystallographic databases.

Diffraction Theory

The main goal of X-ray crystallography is to determine the density of electrons f(**r**) throughout the crystal, where **r** represents the three-dimensional position vector within the crystal. To do this, X-ray scattering is used to collect data about its Fourier transform F(**q**), which is inverted mathematically to obtain the density defined in real space, using the formula

$$f(\mathbf{r}) = \frac{1}{(2\pi)^3} \int F(\mathbf{q}) e^{i\mathbf{q}\cdot\mathbf{r}} d\mathbf{q}$$

where the integral is taken over all values of **q**. The three-dimensional real vector **q** represents a point in reciprocal space, that is, to a particular oscillation in the electron density as one moves in the direction in which **q** points. The length of **q** corresponds to 2π divided by the wavelength of the oscillation. The corresponding formula for a Fourier transform will be used below

$$F(\mathbf{q}) = \int f(\mathbf{r}) e^{-i\mathbf{q}\cdot\mathbf{r}} d\mathbf{r}$$

where the integral is summed over all possible values of the position vector **r** within the crystal.

The Fourier transform F(**q**) is generally a complex number, and therefore has a magnitude |F(**q**)| and a phase φ(**q**) related by the equation

$$F(\mathbf{q}) = |F(\mathbf{q})|\, e^{i\phi(\mathbf{q})}$$

The intensities of the reflections observed in X-ray diffraction give us the magnitudes |F(**q**)| but not the phases φ(**q**). To obtain the phases, full sets of reflections are collected with known alterations to the scattering, either by modulating the wavelength past a certain absorption edge or by adding strongly scattering (i.e., electron-dense) metal atoms such as mercury. Combining the magnitudes and phases yields the full Fourier transform F(**q**), which may be inverted to obtain the electron density f(**r**).

Crystals are often idealized as being perfectly periodic. In that ideal case, the atoms are positioned on a perfect lattice, the electron density is perfectly periodic, and the Fourier transform F(**q**) is zero except when **q** belongs to the reciprocal lattice (the so-called Bragg peaks). In reality, however, crystals are not perfectly periodic; atoms vibrate about their mean position, and there may be disorder of various types, such as mosaicity, dislocations, various point defects, and heterogeneity in the conformation of crystallized molecules. Therefore, the Bragg peaks have a finite width and there may be significant diffuse scattering, a continuum of scattered X-rays that fall between the Bragg peaks.

Intuitive Understanding by Bragg's Law

An intuitive understanding of X-ray diffraction can be obtained from the Bragg model of diffraction. In this model, a given reflection is associated with a set of evenly spaced sheets running through the crystal, usually passing through the centers of the atoms of the crystal lattice. The orientation of a particular set of sheets is identified by its three Miller indices (h, k, l), and let their spacing be noted by d. William Lawrence Bragg proposed a model in which the incoming X-rays are scattered specularly (mirror-like) from each plane; from that assumption, X-rays scattered from adjacent planes will combine constructively (constructive interference) when the angle θ between the plane and the X-ray results in a path-length difference that is an integer multiple n of the X-ray wavelength λ.

$$2d\sin\theta = n\lambda$$

A reflection is said to be indexed when its Miller indices (or, more correctly, its reciprocal lattice vector components) have been identified from the known wavelength and the scattering angle 2θ. Such indexing gives the unit-cell parameters, the lengths and angles of the unit-cell, as well as its space group. Since Bragg's law does not interpret the relative intensities of the reflections, however, it is generally inadequate to solve for the arrangement of atoms within the unit-cell; for that, a Fourier transform method must be carried out.

Scattering as a Fourier Transform

The incoming X-ray beam has a polarization and should be represented as a vector wave; however, for simplicity, let it be represented here as a scalar wave. We also ignore the complication of the time dependence of the wave and just concentrate on the wave's spatial dependence. Plane waves can be represented by a wave vector $\mathbf{k}_{in}$, and so the strength of the incoming wave at time t=0 is given by

$$Ae^{i\mathbf{k}_{in}\cdot\mathbf{r}}$$

At position **r** within the sample, let there be a density of scatterers f(**r**); these scatterers should produce a scattered spherical wave of amplitude proportional to the local amplitude of the incoming wave times the number of scatterers in a small volume dV about **r**

$$\text{amplitude of scattered wave} = Ae^{i\mathbf{k}\cdot\mathbf{r}}Sf(\mathbf{r})dV$$

where S is the proportionality constant.

Let's consider the fraction of scattered waves that leave with an outgoing wave-vector of $\mathbf{k}_{out}$ and strike the screen at $\mathbf{r}_{screen}$. Since no energy is lost (elastic, not inelastic scattering), the wavelengths are the same as are the magnitudes of the wave-vectors $|\mathbf{k}_{in}|=|\mathbf{k}_{out}|$. From the time that the photon is scattered at **r** until it is absorbed at $\mathbf{r}_{screen}$, the photon undergoes a change in phase

$$e^{i\mathbf{k}_{out}\cdot(\mathbf{r}_{screen}-\mathbf{r})}$$

The net radiation arriving at r_{screen} is the sum of all the scattered waves throughout the crystal

$$AS\int d\mathbf{r}f(\mathbf{r})e^{i\mathbf{k}_{in}\cdot\mathbf{r}}e^{i\mathbf{k}_{out}\cdot(\mathbf{r}_{screen}-\mathbf{r})} = ASe^{i\mathbf{k}_{out}\cdot\mathbf{r}_{screen}}\int d\mathbf{r}f(\mathbf{r})e^{i(\mathbf{k}_{in}-\mathbf{k}_{out})\cdot\mathbf{r}}$$

which may be written as a Fourier transform

$$ASe^{i\mathbf{k}_{out}\cdot\mathbf{r}_{screen}}\int d\mathbf{r}f(\mathbf{r})e^{-i\mathbf{q}\cdot\mathbf{r}} = ASe^{i\mathbf{k}_{out}\cdot\mathbf{r}_{screen}}F(\mathbf{q})$$

where $q = k_{out} - k_{in}$. The measured intensity of the reflection will be square of this amplitude

$$A^2S^2|F(\mathbf{q})|^2$$

Friedel and Bijvoet Mates

For every reflection corresponding to a point **q** in the reciprocal space, there is another reflection of the same intensity at the opposite point -**q**. This opposite reflection is known as the Friedel mate of the original reflection. This symmetry results from the mathematical fact that the density of electrons f(**r**) at a position **r** is always a real number. As noted above, f(**r**) is the inverse transform of its Fourier transform F(**q**); however, such an inverse transform is a complex number in general. To ensure that f(**r**) is real, the Fourier transform F(**q**) must be such that the Friedel mates F(−**q**) and F(**q**) are complex conjugates of one another. Thus, F(−**q**) has the same magnitude as F(**q**) but they have the opposite phase, i.e., φ(**q**) = −φ(**q**)

$$F(-\mathbf{q}) = |F(-\mathbf{q})|\,e^{i\phi(-\mathbf{q})} = F^*(\mathbf{q}) = |F(\mathbf{q})|\,e^{-i\phi(\mathbf{q})}$$

The equality of their magnitudes ensures that the Friedel mates have the same intensity $|F|^2$. This symmetry allows one to measure the full Fourier transform from only half the reciprocal space, e.g., by rotating the crystal slightly more than 180° instead of a full 360° revolution. In crystals with significant symmetry, even more reflections may have the same intensity (Bijvoet mates); in such cases, even less of the reciprocal space may need to be measured. In favorable cases of high symmetry, sometimes only 90° or even only 45° of data are required to completely explore the reciprocal space.

The Friedel-mate constraint can be derived from the definition of the inverse Fourier transform

$$f(\mathbf{r}) = \int \frac{d\mathbf{q}}{(2\pi)^3}F(\mathbf{q})e^{i\mathbf{q}\cdot\mathbf{r}} = \int \frac{d\mathbf{q}}{(2\pi)^3}|F(\mathbf{q})|\,e^{i\phi(\mathbf{q})}e^{i\mathbf{q}\cdot\mathbf{r}}$$

Since Euler's formula states that $e^{ix} = \cos(x) + i\sin(x)$, the inverse Fourier transform can be separated into a sum of a purely real part and a purely imaginary part

$$f(\mathbf{r}) = \int \frac{d\mathbf{q}}{(2\pi)^3}|F(\mathbf{q})|\,e^{i(\phi+\mathbf{q}\cdot\mathbf{r})} = \int \frac{d\mathbf{q}}{(2\pi)^3}|F(\mathbf{q})|\cos(\phi+\mathbf{q}\cdot\mathbf{r}) + i\int \frac{d\mathbf{q}}{(2\pi)^3}|F(\mathbf{q})|\sin(\phi+\mathbf{q}\cdot\mathbf{r}) = I_{\cos} + iI_{\sin}$$

The function f(r) is real if and only if the second integral Isin is zero for all values of r. In turn, this is true if and only if the above constraint is satisfied

$$I_{\sin} = \int \frac{d\mathbf{q}}{(2\pi)^3}|F(\mathbf{q})|\sin(\phi+\mathbf{q}\cdot\mathbf{r}) = \int \frac{d\mathbf{q}}{(2\pi)^3}|F(-\mathbf{q})|\sin(-\phi-\mathbf{q}\cdot\mathbf{r}) = -I_{\sin}$$

since $I_{sin} = -I_{sin}$ implies that $I_{sin}=0$.

$$f(\mathbf{r}) = \int \frac{d\mathbf{q}}{(2\pi)^3} F(\mathbf{q}) e^{i\mathbf{q}\cdot\mathbf{r}} = \int \frac{d\mathbf{q}}{(2\pi)^3} |F(\mathbf{q})| \, e^{i\phi(\mathbf{q})} e^{i\mathbf{q}\cdot\mathbf{r}}$$

Since Euler's formula states that $e^{ix} = \cos(x) + i\sin(x)$, the inverse Fourier transform can be separated into a sum of a purely real part and a purely imaginary part

$$f(\mathbf{r}) = \int \frac{d\mathbf{q}}{(2\pi)^3} |F(\mathbf{q})| \, e^{i(\phi+\mathbf{q}\cdot\mathbf{r})} = \int \frac{d\mathbf{q}}{(2\pi)^3} |F(\mathbf{q})| \cos(\phi + \mathbf{q}\cdot\mathbf{r}) + i \int \frac{d\mathbf{q}}{(2\pi)^3} |F(\mathbf{q})| \sin(\phi + \mathbf{q}\cdot\mathbf{r}) = I_{\cos} + iI_{\sin}$$

The function $f(\mathbf{r})$ is real if and only if the second integral $I_{\sin}$ is zero for all values of r. In turn, this is true if and only if the above constraint is satisfied

$$I_{\sin} = \int \frac{d\mathbf{q}}{(2\pi)^3} |F(\mathbf{q})| \sin(\phi + \mathbf{q}\cdot\mathbf{r}) = \int \frac{d\mathbf{q}}{(2\pi)^3} |F(-\mathbf{q})| \sin(-\phi - \mathbf{q}\cdot\mathbf{r}) = -I_{\sin}$$

since $I_{\sin} = -I_{\sin}$ implies that $I_{\sin}=0$.

Ewald's Sphere

Each X-ray diffraction image represents only a slice, a spherical slice of reciprocal space, as may be seen by the Ewald sphere construction. Both $\mathbf{k}_{out}$ and $\mathbf{k}_{in}$ have the same length, due to the elastic scattering, since the wavelength has not changed. Therefore, they may be represented as two radial vectors in a sphere in reciprocal space, which shows the values of $\mathbf{q}$ that are sampled in a given diffraction image. Since there is a slight spread in the incoming wavelengths of the incoming X-ray beam, the values of $|F(\mathbf{q})|$ can be measured only for $\mathbf{q}$ vectors located between the two spheres corresponding to those radii. Therefore, to obtain a full set of Fourier transform data, it is necessary to rotate the crystal through slightly more than 180°, or sometimes less if sufficient symmetry is present. A full 360° rotation is not needed because of a symmetry intrinsic to the Fourier transforms of real functions (such as the electron density), but "slightly more" than 180° is needed to cover all of reciprocal space within a given resolution because of the curvature of the Ewald sphere. In practice, the crystal is rocked by a small amount (0.25-1°) to incorporate reflections near the boundaries of the spherical Ewald's shells.

Patterson Function

A well-known result of Fourier transforms is the autocorrelation theorem, which states that the autocorrelation c(**r**) of a function f(**r**)

$$c(\mathbf{r}) = \int d\mathbf{x} f(\mathbf{x}) f(\mathbf{x}+\mathbf{r}) = \int \frac{d\mathbf{q}}{(2\pi)^3} C(\mathbf{q}) e^{i\mathbf{q}\cdot\mathbf{r}}$$

has a Fourier transform $C(\mathbf{q})$ that is the squared magnitude of $F(\mathbf{q})$

$$C(\mathbf{q}) = |F(\mathbf{q})|^2$$

Therefore, the autocorrelation function c(**r**) of the electron density (also known as the Patterson function) can be computed directly from the reflection intensities, without computing the phases. In principle, this could be used to determine the crystal structure directly; however, it is difficult to realize in practice. The autocorrelation function corresponds to the distribution of vectors between atoms in the crystal; thus, a crystal of N atoms in its unit cell may have N(N-1) peaks in its Patterson function. Given the inevitable errors in measuring the intensities, and the mathematical difficulties of reconstructing atomic positions from the interatomic vectors, this technique is rarely used to solve structures, except for the simplest crystals.

Advantages of a Crystal

In principle, an atomic structure could be determined from applying X-ray scattering to non-crystalline samples, even to a single molecule. However, crystals offer a much stronger signal due to their periodicity. A crystalline sample is by definition periodic; a crystal is composed of many unit cells repeated indefinitely in three independent directions. Such periodic systems have a Fourier transform that is concentrated at periodically repeating points in reciprocal space known as Bragg peaks; the Bragg peaks correspond to the reflection spots observed in the diffraction image. Since the amplitude at these reflections grows linearly with the number N of scatterers, the observed intensity of these spots should grow quadratically, like N^2. In other words, using a crystal concentrates the weak scattering of the individual unit cells into a much more powerful, coherent reflection that can be observed above the noise. This is an example of constructive interference.

In a liquid, powder or amorphous sample, molecules within that sample are in random orientations. Such samples have a continuous Fourier spectrum that uniformly spreads its amplitude thereby reducing the measured signal intensity, as is observed in SAXS. More importantly, the orientational information is lost. Although theoretically possible, it is experimentally difficult to obtain atomic-resolution structures of complicated, asymmetric molecules from such rotationally averaged data. An intermediate case is fiber diffraction in which the subunits are arranged periodically in at least one dimension.

Nobel Prizes for X-ray crystallography

Year	Laureate	Prize	Rationale
1914	Max von Laue	Physics	"For his discovery of the diffraction of X-rays by crystals", an important step in the development of X-ray spectroscopy.
1915	William Henry Bragg	Physics	"For their services in the analysis of crystal structure by means of X-rays",
1915	William Lawrence Bragg	Physics	"For their services in the analysis of crystal structure by means of X-rays",
1962	Max F. Perutz	Chemistry	"for their studies of the structures of globular proteins"
1962	John C. Kendrew	Chemistry	"for their studies of the structures of globular proteins"
1962	James Dewey Watson	Medicine	"For their discoveries concerning the molecular structure of nucleic acids and its significance for information transfer in living material"

Year	Laureate	Prize	Rationale
1962	Francis Harry Compton Crick	Medicine	"For their discoveries concerning the molecular structure of nucleic acids and its significance for information transfer in living material"
1962	Maurice Hugh Frederick Wilkins	Medicine	"For their discoveries concerning the molecular structure of nucleic acids and its significance for information transfer in living material"
1964	Dorothy Hodgkin	Chemistry	"For her determinations by X-ray techniques of the structures of important biochemical substances"
1972	Stanford Moore	Chemistry	"For their contribution to the understanding of the connection between chemical structure and catalytic activity of the active centre of the ribonuclease molecule"
1972	William H. Stein	Chemistry	"For their contribution to the understanding of the connection between chemical structure and catalytic activity of the active centre of the ribonuclease molecule"
1976	William N. Lipscomb	Chemistry	"For his studies on the structure of boranes illuminating problems of chemical bonding"
1985	Jerome Karle	Chemistry	"For their outstanding achievements in developing direct methods for the determination of crystal structures"
1985	Herbert A. Hauptman	Chemistry	"For their outstanding achievements in developing direct methods for the determination of crystal structures"
1988	Johann Deisenhofer	Chemistry	"For their determination of the three-dimensional structure of a photosynthetic reaction centre"
1988	Hartmut Michel	Chemistry	"For their determination of the three-dimensional structure of a photosynthetic reaction centre"
1988	Robert Huber	Chemistry	"For their determination of the three-dimensional structure of a photosynthetic reaction centre"
1997	John E. Walker	Chemistry	"For their elucidation of the enzymatic mechanism underlying the synthesis of adenosine triphosphate (ATP)"
2003	Roderick MacKinnon	Chemistry	"For discoveries concerning channels in cell membranes [...] for structural and mechanistic studies of ion channels"

Year	Laureate	Prize	Rationale
2003	Peter Agre	Chemistry	"For discoveries concerning channels in cell membranes [...] for the discovery of water channels"
2006	Roger D. Kornberg	Chemistry	"For his studies of the molecular basis of eukaryotic transcription"
2009	Ada E. Yonath	Chemistry	"For studies of the structure and function of the ribosome"
2009	Thomas A. Steitz	Chemistry	"For studies of the structure and function of the ribosome"
2009	Venkatraman Ramakrishnan	Chemistry	"For studies of the structure and function of the ribosome"
2012	Brian Kobilka	Chemistry	"For studies of G-protein-coupled receptors"

NUCLEAR MAGNETIC RESONANCE SPECTROSCOPY

Nuclear magnetic resonance spectroscopy, most commonly known as NMR spectroscopy, is a research technique that exploits the magnetic properties of certain atomic nuclei. This type of spectroscopy determines the physical and chemical properties of atoms or the molecules in which they are contained. It relies on the phenomenon of nuclear magnetic resonance and can provide detailed information about the structure, dynamics, reaction state, and chemical environment of molecules. The intramolecular magnetic field around an atom in a molecule changes the resonance frequency, thus giving access to details of the electronic structure of a molecule and its individual functional groups.

Most frequently, NMR spectroscopy is used by chemists and biochemists to investigate the properties of organic molecules, although it is applicable to any kind of sample that contains nuclei possessing spin. Suitable samples range from small compounds analyzed with 1-dimensional proton or carbon-13 NMR spectroscopy to large proteins or nucleic acids using 3 or 4-dimensional techniques. The impact of NMR spectroscopy on the sciences has been substantial because of the range of information and the diversity of samples, including solutions and solids.

NMR spectra are unique, well-resolved, analytically tractable and often highly predictable for small molecules. Thus, in organic chemistry practice, NMR analysis is used to confirm the identity of a substance. Different functional groups are obviously distinguishable, and identical functional groups with differing neighboring substituents still give distinguishable signals. NMR has largely replaced traditional wet chemistry tests such as color reagents or typical chromatography for identification. A disadvantage is that a relatively large amount, 2–50 mg, of a purified substance is required, although it may be recovered through a workup. Preferably, the sample should be dissolved in a solvent, because NMR analysis of solids requires a dedicated MAS machine and may not give equally well-resolved spectra. The timescale of NMR is relatively long, and thus it is not suitable for observing fast phenomena, producing only an averaged spectrum.

Although large amounts of impurities do show on an NMR spectrum, better methods exist for detecting impurities, as NMR is inherently not very sensitive - though at higher frequencies, sensitivity narrows.

NMR spectrometers are relatively expensive; universities usually have them, but they are less common in private companies. Modern NMR spectrometers have a very strong, large and expensive liquid helium-cooled superconducting magnet, because resolution directly depends on magnetic field strength. Less expensive machines using permanent magnets and lower resolution are also available, which still give sufficient performance for certain application such as reaction monitoring and quick checking of samples. There are even benchtop NMR spectrometers.

A 900MHz NMR instrument with a 21.1 T magnet at HWB-NMR, Birmingham, UK

The Purcell group at Harvard University and the Bloch group at Stanford University independently developed NMR spectroscopy in the late 1940s and early 1950s. Edward Mills Purcell and Felix Bloch shared the 1952 Nobel Prize in Physics for their discoveries.

Basic NMR Techniques

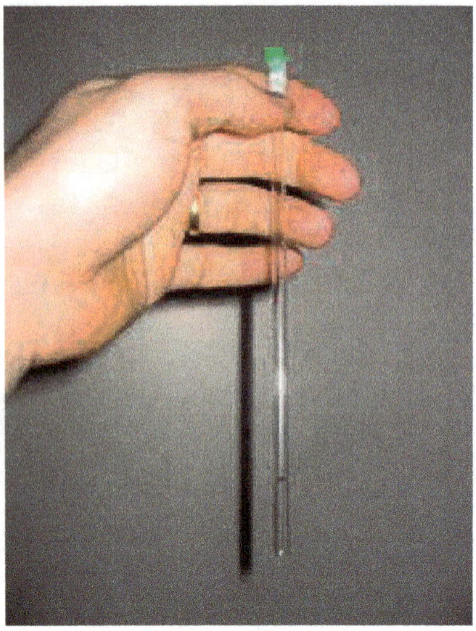

The NMR sample is prepared in a thin-walled glass tube - an NMR tube.

Resonant Frequency

When placed in a magnetic field, NMR active nuclei (such as ^{1}H or ^{13}C) absorb electromagnetic radiation at a frequency characteristic of the isotope. The resonant frequency, energy of the absorption, and the intensity of the signal are proportional to the strength of the magnetic field. For example, in a 21 Tesla magnetic field, protons resonate at 900 MHz. It is common to refer to a 21 T magnet as a 900 MHz magnet, although different nuclei resonate at a different frequency at this field strength in proportion to their nuclear magnetic moments.

Sample Handling

A NMR spectrometer typically consists of a spinning sample-holder inside a very strong magnet, a radio-frequency emitter and a receiver with a probe (an antenna assembly) that goes inside the magnet to surround the sample, optionally gradient coils for diffusion measurements and electronics to control the system. Spinning the sample is necessary to average out diffusional motion. Whereas, measurements of diffusion constants (diffusion ordered spectroscopy or DOSY) are done the sample stationary and spinning off, and flow cells can be used for online analysis of process flows.

Deuterated Solvents

The vast majority of nuclei in a solution would belong to the solvent, and most regular solvents are hydrocarbons and would contain NMR-reactive protons. Thus, deuterium (hydrogen-2) is substituted (99+%). The most used deuterated solvent is deuterochloroform

($CDCl_3$), although deuterium oxide (D_2O) and deuterated DMSO (DMSO-d_6) are used for hydrophilic analytes. The chemical shifts are slightly different in different solvents, depending on electronic solvation effects. NMR spectra are often calibrated against the known solvent residual proton peak instead of added tetramethylsilane.

Shim and Lock

To detect the very small frequency shifts due to nuclear magnetic resonance, the applied magnetic field must be constant throughout the sample volume. High resolution NMR spectrometers use shims to adjust the homogeneity of the magnetic field to parts per billion (ppb) in a volume of a few cubic centimeters. In order to detect and compensate for inhomogeneity and drift in the magnetic field, the spectrometer maintains a "lock" on the solvent deuterium frequency with a separate lock unit. The operator usually has to optimize the shim parameters manually to obtain the best possible intensity and prevent artifacts.

Acquisition of Spectra

Upon excitation of the sample with a radio frequency (60–1000 MHz) pulse, a nuclear magnetic resonance response - a free induction decay (FID) - is obtained. It is a very weak signal, and requires sensitive radio receivers to pick up. A Fourier transform is carried out to extract the frequency-domain spectrum from the raw time-domain FID. A spectrum from a single FID has a low signal-to-noise ratio, but fortunately it improves readily with averaging of repeated acquisitions. Good ^{1}H NMR spectra can be acquired with 16 repeats, which takes only minutes. However, for elements heavier than hydrogen, the relaxation time is rather long, e.g. around 8 seconds for ^{13}C. Thus, acquisition of quantitative heavy-element spectra can be time-consuming, taking tens of minutes to hours.

If the second excitation pulse is sent prematurely before the relaxation is complete, the average magnetization vector still points in a nonparallel direction, giving suboptimal absorption and emission of the pulse. In practice, the peak areas are then not proportional to the stoichiometry; only the presence, but not the amount of functional groups is possible to discern. An inversion recovery experiment can be done to determine the relaxation time and thus the required delay between pulses. A 180° pulse, an adjustable delay, and a 90° pulse is transmitted. When the 90° pulse exactly cancels out the signal, the delay corresponds to the time needed for 90° of relaxation. Inversion recovery is worthwhile for quantitive ^{13}C, 2D and other time-consuming experiments.

Chemical Shift

A spinning charge generates a magnetic field that results in a magnetic moment proportional to the spin. In the presence of an external magnetic field, two spin states exist (for a spin 1/2 nucleus): one spin up and one spin down, where one aligns with the magnetic field and the other opposes it. The difference in energy (ΔE) between the two spin states increases as the strength of the field increases, but this difference is usually very small, leading to the requirement for strong NMR magnets (1-20 T for modern NMR instruments). Irradiation of the sample with energy corresponding to the exact spin state separation of a specific set of nuclei will cause excitation of those set of nuclei in the lower energy state to the higher energy state.

For spin 1/2 nuclei, the energy difference between the two spin states at a given magnetic field strength is proportional to their magnetic moment. However, even if all protons have the same magnetic moments, they do not give resonant signals at the same frequency values. This difference arises from the differing electronic environments of the nucleus of interest. Upon application of an external magnetic field, these electrons move in response to the field and generate local magnetic fields that oppose the much stronger applied field. This local field thus "shields" the proton from the applied magnetic field, which must therefore be increased in order to achieve resonance (absorption of rf energy). Such increments are very small, usually in parts per million (ppm). For instance, the proton peak from an aldehyde is shifted ca. 10 ppm compared to a hydrocarbon peak, since as an electron-withdrawing group, the carbonyl deshields the proton by reducing the local electron density. The difference between 2.3487 T and 2.3488 T is therefore about 42 ppm. However a frequency scale is commonly used to designate the NMR signals, even though the spectrometer may operate by sweeping the magnetic field, and thus the 42 ppm is 4200 Hz for a 100 MHz reference frequency (rf).

However given that the location of different NMR signals is dependent on the external magnetic field strength and the reference frequency, the signals are usually reported relative to a reference signal, usually that of TMS (tetramethylsilane). Additionally, since the distribution of NMR signals is field dependent, these frequencies are divided by the spectrometer frequency. However, since we are dividing Hz by MHz, the resulting number would be too small, and thus it is multiplied by a million. This operation therefore gives a locator number called the "chemical shift" with units of parts per million. In general, chemical shifts for protons are highly predictable since the shifts are primarily determined by simpler shielding effects (electron density), but the chemical shifts for many heavier nuclei are more strongly influenced by other factors including excited states ("paramagnetic" contribution to shielding tensor).

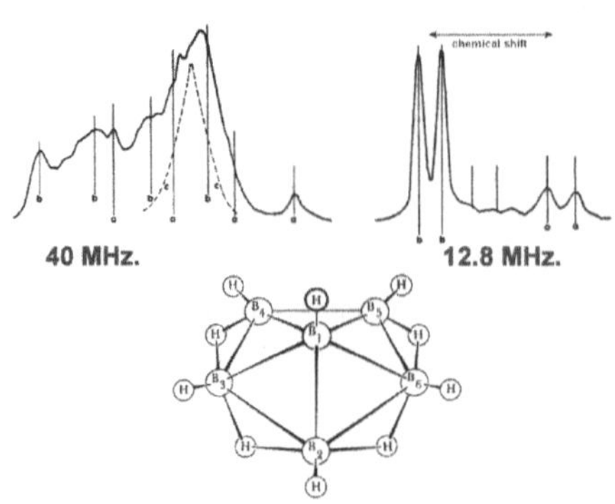

Example of the chemical shift: NMR spectrum of hexaborane B_6H_{10}showing peaks shifted in frequency, which give clues as to the molecular structure. (click to read interpretation details)

The chemical shift provides information about the structure of the molecule. The conversion of the raw data to this information is called assigning the spectrum. For example, for the ^{1}H-NMR spectrum for ethanol (CH_3CH_2OH), one would expect signals at each of three specific chemical shifts: one for the CH_3 group, one for the CH_2 group and one for the OH group. A typical CH_3 group has a shift around 1 ppm, a CH_2 attached to an OH has a shift of around 4 ppm and an OH has a shift anywhere from 2–6 ppm depending on the solvent used and the amount of hydrogen bonding. While the O atom does draw electron density away from the attached H through their mutual sigma bond, the electron lone pairs on the O bathe the H in their shielding effect.

In Paramagnetic NMR spectroscopy, measurements are conducted on paramagnetic samples. The paramagnetism gives rise to very diverse chemical shifts. In 1H NMR spectroscopy, the chemical shift range can span 500 ppm.

Because of molecular motion at room temperature, the three methyl protons average out during the NMR experiment (which typically requires a few ms). These protons become degenerate and form a peak at the same chemical shift.

The shape and area of peaks are indicators of chemical structure too. In the example above—the proton spectrum of ethanol—the CH_3 peak has three times the area as the OH peak. Similarly the CH_2 peak would be twice the area of the OH peak but only 2/3 the area of the CH_3 peak.

Software allows analysis of signal intensity of peaks, which under conditions of optimal relaxation, correlate with the number of protons of that type. Analysis of signal intensity is done by integration—the mathematical process that calculates the area under a curve. The analyst must integrate the peak and not measure its height because the peaks also have width—and thus its size is dependent on its area not its height. However, it should be mentioned that the number of protons, or any other observed nucleus, is only proportional to the intensity, or the integral, of the NMR signal in the very simplest one-dimensional NMR experiments. In more elaborate experiments, for instance, experiments typically used to obtain carbon-13 NMR spectra, the integral of the signals depends on the relaxation rate of the nucleus, and its scalar and dipolar coupling constants. Very often these factors are poorly known - therefore, the integral of the NMR signal is very difficult to interpret in more complicated NMR experiments.

J-coupling

Some of the most useful information for structure determination in a one-dimensional NMR spectrum comes from **J-coupling** or **scalar coupling** (a special case of spin-spin coupling) between NMR active nuclei. This coupling arises from the interaction of different spin states through the chemical bonds of a molecule and results in the splitting of NMR signals. These splitting patterns can be complex or simple and, likewise, can be straightforwardly interpretable or deceptive. This coupling provides detailed insight into the connectivity of atoms in a molecule.

Coupling to n equivalent (spin ½) nuclei splits the signal into a n+1 **multiplet** with intensity ratios following Pascal's triangle as described on the right. Coupling to additional spins will lead to further splittings of each component of the multiplet e.g. coupling to two different spin ½ nuclei with significantly different coupling constants will lead to a doublet of doublets (abbreviation: dd). Note that coupling between nuclei that are chemically equivalent (that is, have the same chemical shift) has no effect on the NMR spectra and couplings between nuclei that are distant (usually more than 3 bonds apart for protons in flexible molecules) are usually too small to cause observable splittings. Long-range couplings over more than three bonds can often be observed in cyclic and aromatic compounds, leading to more complex splitting patterns.

For example, in the proton spectrum for ethanol described above, the CH_3 group is split into a triplet with an intensity ratio of 1:2:1 by the two neighboring CH_2 protons. Similarly, the CH_2 is split into a quartet with an intensity ratio of 1:3:3:1 by the three neighboring CH_3 protons. In principle, the two CH_2 protons would also be split again into a doublet to form a doublet of quartets by the hydroxyl proton, but intermolecular exchange of the acidic hydroxyl proton often results in a loss of coupling information.

Coupling to any spin ½ nuclei such as phosphorus-31 or fluorine-19 works in this fashion (although the magnitudes of the coupling constants may be very different). But the splitting patterns differ from those described above for nuclei with spin greater than ½ because the spin quantum number has more than two possible values. For instance, coupling to deuterium (a spin 1 nucleus) splits the signal into a 1:1:1 triplet because the spin 1 has three spin states. Similarly, a spin 3/2 nucleus splits a signal into a 1:1:1:1 quartet and so on.

Coupling combined with the chemical shift (and the integration for protons) tells us not only about the chemical environment of the nuclei, but also the number of neighboring NMR active nuclei within the molecule. In more complex spectra with multiple peaks at similar chemical shifts or in spectra of nuclei other than hydrogen, coupling is often the only way to distinguish different nuclei.

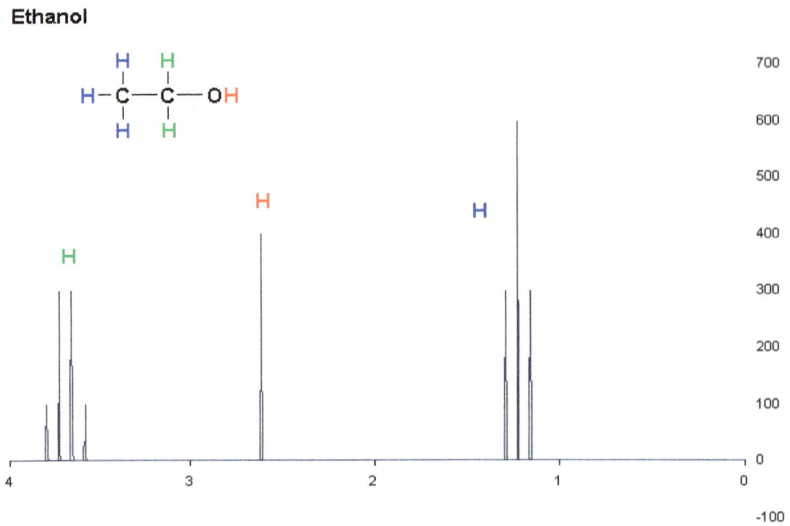

Example ^1H NMR spectrum (1-dimensional) of ethanol plotted as signal intensity vs. chemical shift. There are three different types of H atoms in ethanol regarding NMR. The hydrogen (H) on the -OH group is not coupling with the other H atoms and appears as a singlet, but the CH_3- and the -CH_2- hydrogens are coupling with each other, resulting in a triplet and quartet respectively.

Second-Order (or Strong) Coupling

The above description assumes that the coupling constant is small in comparison with the difference in NMR frequencies between the inequivalent spins. If the shift separation decreases (or the coupling strength increases), the multiplet intensity patterns are first distorted, and then become more complex and less easily analyzed (especially if more than two spins are involved). Intensification of some peaks in a multiplet is achieved at the expense of the remainder, which sometimes almost disappear in the background noise, although the integrated area under the peaks remains constant. In most high-field NMR, however, the distortions are usually modest and the characteristic distortions (roofing) can in fact help to identify related peaks.

Some of these patterns can be analyzed with the method published by John Pople, though it has limited scope.

Second-order effects decrease as the frequency difference between multiplets increases, so that high-field (i.e. high-frequency) NMR spectra display less distortion than lower frequency spectra. Early spectra at 60 MHz were more prone to distortion than spectra from later machines typically operating at frequencies at 200 MHz or above.

Magnetic Inequivalence

More subtle effects can occur if chemically equivalent spins (i.e., nuclei related by symmetry and so having the same NMR frequency) have different coupling relationships to external spins. Spins that are chemically equivalent but are not indistinguishable (based on their coupling relationships) are termed magnetically inequivalent. For example, the 4 H sites of 1,2-dichlorobenzene divide into two chemically equivalent pairs by symmetry, but an individual member of one of the pairs has different couplings to the spins making up the other pair. Magnetic inequivalence can lead to highly complex spectra which can only be analyzed by computational modeling. Such effects are more common in NMR spectra of aromatic and other non-flexible systems, while conformational averaging about C-C bonds in flexible molecules tends to equalize the couplings between protons on adjacent carbons, reducing problems with magnetic inequivalence.

Correlation Spectroscopy

Correlation spectroscopy is one of several types of two-dimensional nuclear magnetic resonance (NMR) spectroscopy or 2D-NMR. This type of NMR experiment is best known by its acronym, COSY. Other types of two-dimensional NMR include J-spectroscopy, exchange spectroscopy (EXSY), Nuclear Overhauser effect spectroscopy (NOESY), total correlation spectroscopy (TOCSY) and heteronuclear correlation experiments, such as HSQC, HMQC, and HMBC. In correlation spectroscopy, emission is centered on the peak of an individual nucleus; if its magnetic field is correlated with another nucleus by through-bond (COSY, HSQC, etc.) or through-space (NOE) coupling, a response can also be detected on the frequency of the correlated nucleus. Two-dimensional NMR spectra provide more information about a molecule than one-dimensional NMR spectra and are especially useful in determining the structure of a molecule, particularly for molecules that are too complicated to work with using one-dimensional NMR. The first two-dimensional experiment, COSY, was proposed by Jean Jeener, a professor at Université Libre de Bruxelles, in 1971. This experiment was later implemented by Walter P. Aue, Enrico Bartholdi and Richard R. Ernst, who published their work in 1976.

Solid-State Nuclear Magnetic Resonance

A variety of physical circumstances do not allow molecules to be studied in solution, and at the same time not by other spectroscopic techniques to an atomic level, either. In solid-phase media, such as crystals, microcrystalline powders, gels, anisotropic solutions, etc., it is in particular the dipolar coupling and chemical shift anisotropy that become dominant to the behaviour of the nuclear spin systems. In conventional solution-state NMR spectroscopy, these additional interactions would lead to a significant broadening of spectral lines. A variety of techniques allows establishing high-resolution conditions, that can, at least for ^{13}C spectra, be comparable to solution-state NMR spectra.

Two important concepts for high-resolution solid-state NMR spectroscopy are the limitation of possible molecular orientation by sample orientation, and the reduction of anisotropic nuclear magnetic interactions by sample spinning. Of the latter approach, fast spinning around the magic angle is a very prominent method, when the system comprises spin 1/2 nuclei. Spinning rates of ca. 20 kHz are used, which demands special equipment. A number of intermediate techniques, with samples of partial alignment or reduced mobility, is currently being used in NMR spectroscopy.

Applications in which solid-state NMR effects occur are often related to structure investigations on membrane proteins, protein fibrils or all kinds of polymers, and chemical analysis in inorganic chemistry, but also include "exotic" applications like the plant leaves and fuel cells. For example, Rahmani et al. studied the effect of pressure and temperature on the bicellar structures' self-assembly using deuterium NMR spectroscopy.

Biomolecular NMR Spectroscopy

Proteins

Much of the innovation within NMR spectroscopy has been within the field of protein NMR spectroscopy, an important technique in structural biology. A common goal of these investigations is to obtain high resolution 3-dimensional structures of the protein, similar to what can be achieved by X-ray crystallography. In contrast to X-ray crystallography, NMR spectroscopy is usually limited to proteins smaller than 35 kDa, although larger structures have been solved. NMR spectroscopy is often the only way to obtain high resolution information on partially or wholly intrinsically unstructured proteins. It is now a common tool for the determination of Conformation Activity Relationships where the structure before and after interaction with, for example, a drug candidate is compared to its known biochemical activity. Proteins are orders of magnitude larger than the small organic molecules discussed earlier in this article, but the basic NMR techniques and some NMR theory also applies. Because of the much higher number of atoms present in a protein molecule in comparison with a small organic compound, the basic 1D spectra become crowded with overlapping signals to an extent where direct spectral analysis becomes untenable. Therefore, multidimensional (2, 3 or 4D) experiments have been devised to deal with this problem. To facilitate these experiments, it is desirable to isotopically label the protein with ^{13}C and ^{15}N because the predominant naturally occurring isotope ^{12}C is not NMR-active and the nuclear quadrupole moment of the predominant naturally occurring ^{14}N isotope prevents high resolution information from being obtained from this nitrogen isotope. The most important method used for structure determination of proteins utilizes NOE experiments to measure distances between pairs of atoms within the molecule. Subsequently, the distances obtained are used to generate a 3D structure of the molecule by solving a distance geometry problem. NMR can also be used to obtain information on the dynamics and conformational flexibility of different regions of a protein.

Nucleic Acids

"Nucleic acid NMR" is the use of NMR spectroscopy to obtain information about the structure and dynamics of polynucleic acids, such as DNA or RNA. As of 2003, nearly half of all known RNA structures had been determined by NMR spectroscopy.

Nucleic acid and protein NMR spectroscopy are similar but differences exist. Nucleic acids have a smaller percentage of hydrogen atoms, which are the atoms usually observed in NMR spectroscopy, and because nucleic acid double helices are stiff and roughly linear, they do not fold back on themselves to give "long-range" correlations. The types of NMR usually done with nucleic acids are ^{1}H or proton NMR, ^{13}C NMR, ^{15}N NMR, and ^{31}P NMR. Two-dimensional NMR methods are almost always used, such as correlation spectroscopy (COSY) and total coherence transfer spectroscopy (TOCSY) to detect through-bond nuclear couplings, and nuclear Overhauser effect spectroscopy (NOESY) to detect couplings between nuclei that are close to each other in space.

Parameters taken from the spectrum, mainly NOESY cross-peaks and coupling constants, can be used to determine local structural features such as glycosidic bond angles, dihedral angles (using the Karplus equation), and sugar pucker conformations. For large-scale structure, these local parameters must be supplemented with other structural assumptions or models, because errors add up as the double helix is traversed, and unlike with proteins, the double helix does not have a compact interior and does not fold back upon itself. NMR is also useful for investigating nonstandard geometries such as bent helices, non-Watson–Crick basepairing, and coaxial stacking. It has been especially useful in probing the structure of natural RNA oligonucleotides, which tend to adopt complex conformations such as stem-loops and pseudoknots. NMR is also useful for probing the binding of nucleic acid molecules to other molecules, such as proteins or drugs, by seeing which resonances are shifted upon binding of the other molecule.

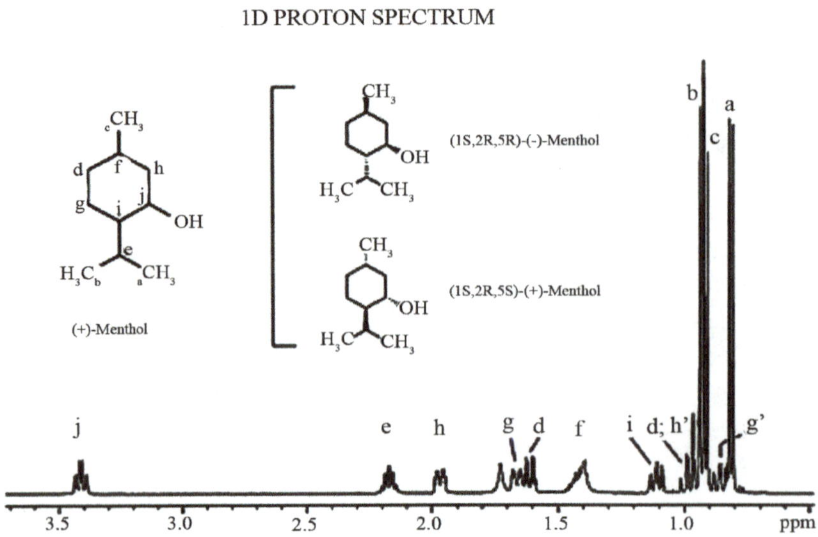

^{1}H NMR spectrum of menthol with chemical shift in ppm on the horizontal axis. Each magnetically inequivalent proton has a characteristic shift, and couplings to other protons appear as splitting of the peaks into multiplets: e.g. peak a, because of the three magnetically equivalent protons in methyl group a, couple to one adjacent proton (e) and thus appears as a doublet.

ATOMIC-FORCE MICROSCOPY

Atomic-force microscopy (AFM) or scanning-force Microscopy (SFM) is a very-high-resolution type of scanning probe microscopy (SPM), with demonstrated resolution on the order of fractions of a nanometer, more than 1000 times better than the optical diffraction limit.

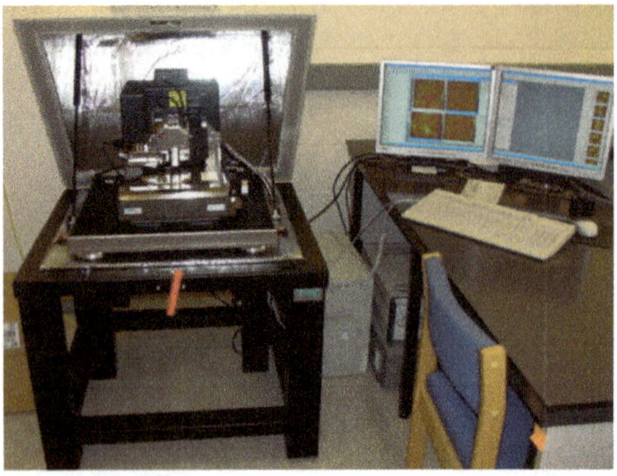

An atomic-force microscope on the left with controlling computer on the right.

Overview

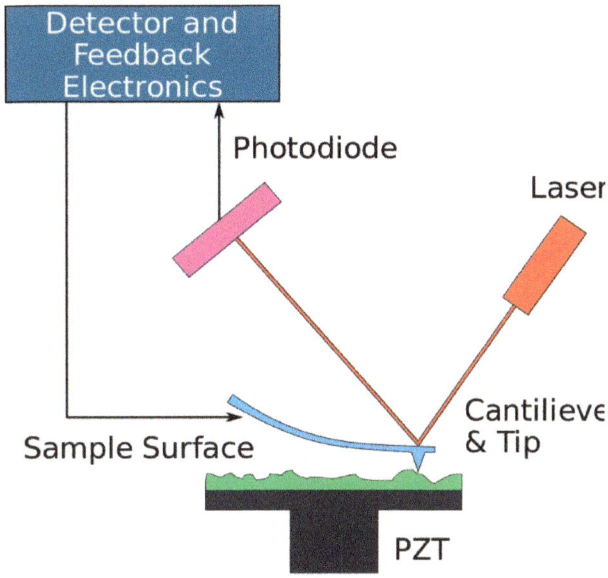

Block diagram of atomic-force microscope using beam deflection detection. As the cantilever is displaced via its interaction with the surface, so too will the reflection of the laser beam be displaced on the surface of the photodiode.

Atomic-force microscopy (AFM) or scanning-force microscopy (SFM) is a type of scanning probe microscopy (SPM), with demonstrated resolution on the order of fractions of a nanometer, more than 1000 times better than the optical diffraction limit. The information is gathered by "feeling" or "touching" the surface with a mechanical probe. Piezoelectric elements that facilitate tiny but accurate and precise movements on (electronic) command enable very precise scanning.

Abilities

The AFM has three major abilities: force measurement, imaging, and manipulation.

In force measurement, AFMs can be used to measure the forces between the probe and the sample as a function of their mutual separation. This can be applied to perform force spectroscopy.

For imaging, the reaction of the probe to the forces that the sample imposes on it can be used to form an image of the three-dimensional shape (topography) of a sample surface at a high resolution. This is achieved by raster scanning the position of the sample with respect to the tip and recording the height of the probe that corresponds to a constant probe-sample interaction (see section topographic imaging in AFM for more details). The surface topography is commonly displayed as a pseudocolor plot. In manipulation, the forces between tip and sample can also be used to change the properties of the sample in a controlled way. Examples of this include atomic manipulation, scanning probe lithography and local stimulation of cells. Simultaneous with the acquisition of topographical images, other properties of the sample can be measured locally and displayed as an image, often with similarly high resolution. Examples of such properties are mechanical properties like stiffness or adhesion strength and electrical properties such as conductivity or surface potential. In fact, the majority of SPM techniques are extensions of AFM that use this modality.

Other Microscopy Technologies

Compared to competitive technologies such as optical microscopy and electron microscopy, the major difference between these and the atomic-force microscope is that the latter does not use lenses or beam irradiation. Therefore, it does not suffer from a limitation of space resolution due to diffraction limit and aberration, and it is not necessary to prepare a space for guiding the beam (by creating a vacuum) or to stain the sample.

There are several types of scanning microscopy including scanning probe microscopy (which includes AFM, STM and near-field scanning optical microscope (SNOM/NSOM), STED microscopy (STED), and scanning electron microscopy). Although SNOM and STED use visible light to illuminate the sample, their resolution is not constrained by the diffraction limit.

Configuration

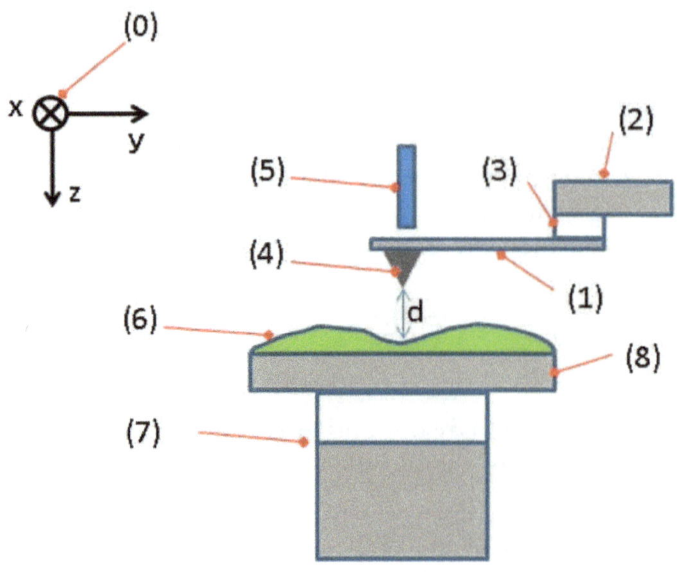

Typical configuration of an AFM.
(1):: Cantilever, **(2)**: Support for cantilever, **(3)**: Piezoelectric element(to oscillate cantilever at its eigen frequency.), **(4)**: Tip (Fixed to open end of a cantilever, acts as the probe), **(5)**: Detector of deflection and motion of the cantilever, **(6)**: Sample to be measured by AFM, **(7)**: xyz drive, (moves sample (6) and stage (8) in x, y, and z directions with respect to a tip apex (4)), and **(8)**: Stage.

The small spring-like cantilever (1) is carried by the support (2). Optionally, a piezoelectric element (3) oscillates the cantilever (1). The sharp tip (4) is fixed to the free end of the cantilever (1). The detector (5) records the deflection and motion of the cantilever (1). The sample (6) is mounted on the sample stage (8). An xyz drive (7) permits to displace the sample (6) and the sample stage (8) in x, y, and z directions with respect to the tip apex (4). Although the drive attached to the sample, the drive can also be attached to the tip, or independent drives can be attached to both, since it is the relative displacement of the sample and tip that needs to be controlled. Controllers and plotter are Numbers in parentheses correspond to numbered features in Coordinate directions are defined by the coordinate system (0).

According to the configuration described above, the interaction between tip and sample, which can be an atomic scale phenomenon, is transduced into changes of the motion of cantilever which is a macro scale phenomenon. Several different aspects of the cantilever motion can be used to quantify the interaction between the tip and sample, most commonly the value of the deflection, the amplitude of an imposed oscillation of the cantilever, or the shift in resonance frequency of the cantilever (see section Imaging Modes).

Detector

The detector (5) of AFM measures the deflection (displacement with respect to the equilibrium position) of the cantilever and converts it into an electrical signal. The intensity of this signal will be proportional to the displacement of the cantilever.

Various methods of detection can be used, e.g. interferometry, optical levers, the piezoresistive method, the piezoelectric method, and STM-based detectors (see section "AFM cantilever deflection measurement".).

Image Formation

Note: This paragraphs assumes the 'contact mode' is used (see section Imaging Modes). For other imaging modes, the process is similar, except that 'deflection' should be replaced by the appropriate feedback variable.

When using the AFM to image a sample, the tip is brought into contact with the sample, and the sample is raster scanned along an x-y grid (fig 4). Most commonly, an electronic feedback loop is employed to keep the probe-sample force constant during scanning. This feedback loop has the cantilever deflection as input, and its output controls the distance along the z axis between the probe support and the sample support. As long as the tip remains in contact with the sample, and the sample is scanned in the x-y plane, height variations in the sample will change the deflection of the cantilever. The feedback then adjusts the height of the probe support so that the deflection is restored to a user-defined value (the setpoint). A properly adjusted feedback loop adjusts the support-sample separation continuously during the scanning motion, such that the deflection remains approximately constant. In this situation, the feedback output equals the sample surface topography to within a small error.

Historically, a different operation method has been used, in which the sample-probe support distance is kept constant and not controlled by a feedback (servo mechanism). In this mode, usually referred to as 'constant height mode', the deflection of the cantilever is recorded as a function of the sample x-y position. As long as the tip is in contact with the sample, the deflection then corresponds to surface topography. The main reason this method is not very popular anymore, is that the forces between tip and sample are not controlled, which can lead to forces high enough to damage the tip or the sample. It is however common practice to record the deflection even when scanning in 'constant force mode', with feedback. This reveals the small tracking error of the feedback, and can sometimes reveal features that the feedback was not able to adjust for.

The AFM signals, such as sample height or cantilever deflection, are recorded on a computer during the x-y scan. They are plotted in a pseudocolor image, in which each pixel represents an x-y position on the sample, and the color represents the recorded signal.

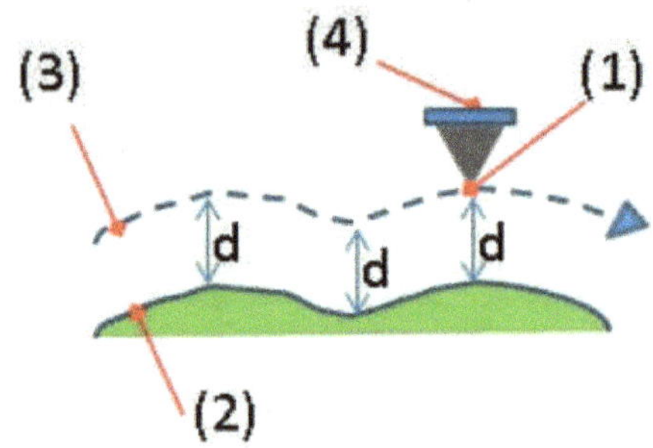

Topographic image forming by AFM.**(1)**: Tip apex, **(2)**: Sample surface, **(3)**: Z-orbit of Tip apex, **(4)**: Cantilever.

AFM was invented by IBM Scientists in 1982.The precursor to the AFM, the scanning tunneling microscope (STM), was developed by Gerd Binnig and Heinrich Rohrer in the early 1980s at IBM Research - Zurich, a development that earned them the Nobel Prize for Physics in 1986. Binnig invented the atomic-force microscope and the first experimental implementation was made by Binnig, Quate and Gerber in 1986.

The first commercially available atomic-force microscope was introduced in 1989. The AFM is one of the foremost tools for imaging, measuring, and manipulating matter at the nanoscale.

Applications

The AFM has been applied to problems in a wide range of disciplines of the natural sciences, including solid state physics, semiconductor science and technology, molecular engineering, polymer chemistry and physics, surface chemistry, molecular biology, cell biology and medicine.

Applications in the field of solid state physics include (a) the identification of atoms at a surface, (b) the evaluation of interactions between a specific atom and its neighboring atoms, and (c) the study of changes in physical properties arising from changes in an atomic arrangement through atomic manipulation.

In cellular biology, AFM can be used to (a) attempt to distinguish cancer cells and normal cells based on a hardness of cells, and (b) to evaluate interactions between a specific cell and its neighboring cells in a competitive culture system.

In some variations, electric potentials can also be scanned using conducting cantilevers. In more advanced versions, currents can be passed through the tip to probe the electrical conductivity or transport of the underlying surface, but this is a challenging task with few research groups reporting consistent data (as of 2004).

Principles

Electron micrograph of a used AFM cantilever. Image width ~100 micrometers

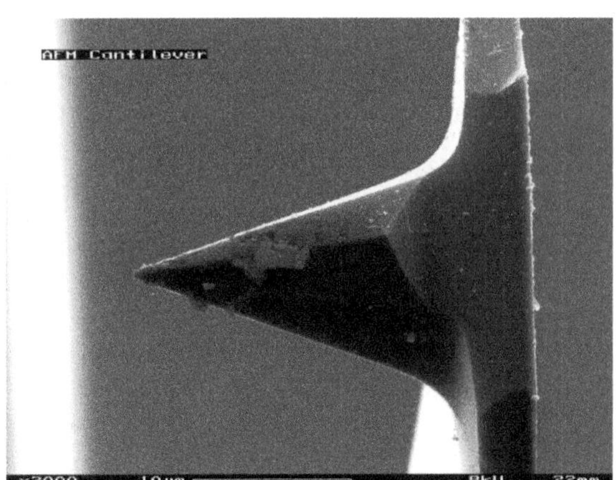

Electron micrograph of a used AFM cantilever. Image width ~30 micrometers

The AFM consists of a cantilever with a sharp tip (probe) at its end that is used to scan the specimen surface. The cantilever is typically silicon or silicon nitride with a tip radius of curvature on the order of nanometers. When the tip is brought into proximity of a sample surface, forces between the tip and the sample lead to a deflection of the cantilever according to Hooke's law. Depending on the situation, forces that are measured in AFM include mechanical contact force, van der Waals forces, capillary forces, chemical bonding, electrostatic forces, magnetic forces (see magnetic force microscope, MFM), Casimir forces, solvation forces, etc. Along with force, additional quantities may simultaneously be measured through the use of specialized types of probes (see scanning thermal microscopy, scanning joule expansion microscopy, photothermal microspectroscopy, etc.).

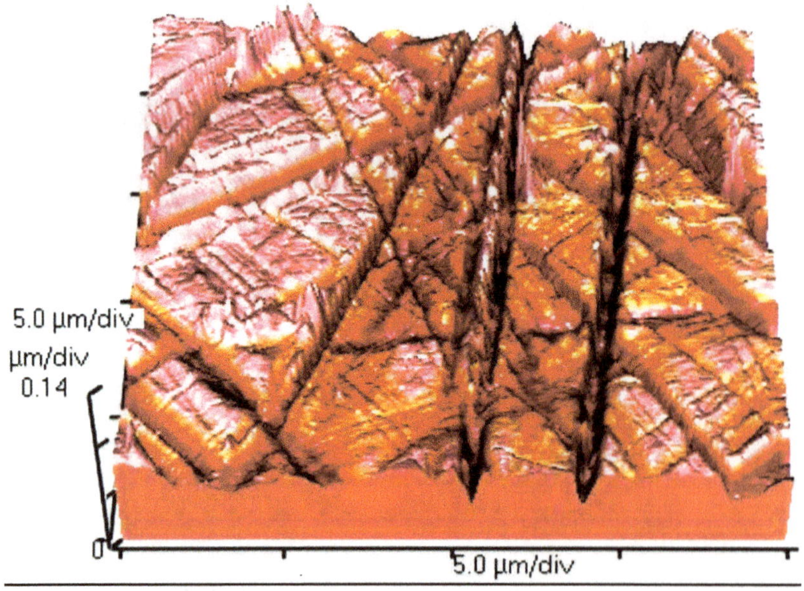

Atomic-force microscope topographical scan of a glass surface. The micro and nano-scale features of the glass can be observed, portraying the roughness of the material. The image space is (x,y,z) = (20 µm × 20 µm × 420 nm).

The AFM can be operated in a number of modes, depending on the application. In general, possible imaging modes are divided into static (also called contact) modes and a variety of dynamic (non-contact or "tapping") modes where the cantilever is vibrated or oscillated at a given frequency.

Imaging Modes

AFM operation is usually described as one of three modes, according to the nature of the tip motion: contact mode, also called static mode (as opposed to the other two modes, which are called dynamic modes); tapping mode, also called intermittent contact, AC mode, or vibrating mode, or, after the detection mechanism, amplitude modulation AFM; non-contact mode, or, again after the detection mechanism, frequency modulation AFM. It should be noted that despite the nomenclature, repulsive contact can occur or be avoided both in amplitude modulation AFM and frequency modulation AFM, depending on the settings.

Contact Mode

In contact mode, the tip is "dragged" across the surface of the sample and the contours of the surface are measured either using the deflection of the cantilever directly or, more commonly, using the feedback signal required to keep the cantilever at a constant position. Because the measurement of a static signal is prone to noise and drift, low stiffness cantilevers (i.e. cantilevers with a low spring constant, k) are used to achieve a large enough deflection signal while keeping the interaction force low. Close to the surface of the sample, attractive forces can be quite strong, causing the tip to "snap-in" to the surface. Thus, contact mode AFM is almost always done at a depth where the overall force is repulsive, that is, in firm "contact" with the solid surface.

Tapping Mode

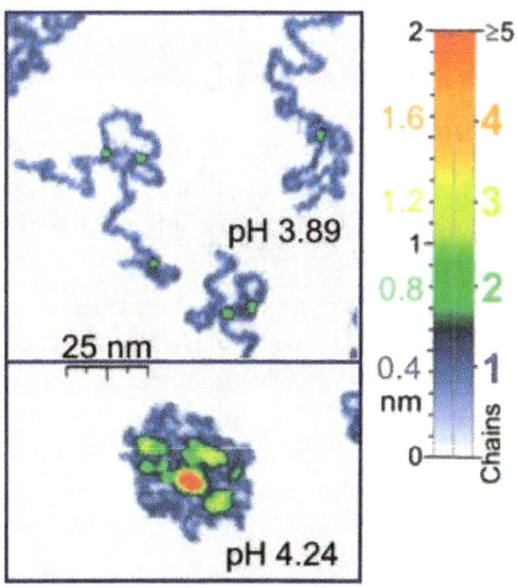

Single polymer chains (0.4 nm thick) recorded in a tapping mode under aqueous media with different pH.

In ambient conditions, most samples develop a liquid meniscus layer. Because of this, keeping the probe tip close enough to the sample for short-range forces to become detectable while preventing the tip from sticking to the surface presents a major problem for non-contact dynamic mode in ambient conditions. Dynamic contact mode (also called intermittent contact, AC mode or tapping mode) was developed to bypass this problem. Nowadays, tapping mode is the most frequently used AFM mode when operating in ambient conditions or in liquids.

In tapping mode, the cantilever is driven to oscillate up and down at or near its resonance frequency. This oscillation is commonly achieved with a small piezo element in the cantilever holder, but other possibilities include an AC magnetic field (with magnetic cantilevers), piezoelectric cantilevers, or periodic heating with a modulated laser beam. The amplitude of this oscillation usually varies from several nm to 200 nm. In tapping mode, the frequency and amplitude of the driving signal are kept constant, leading to a constant amplitude of the cantilever oscillation as long as there is no drift or interaction with the surface. The interaction of forces acting on the cantilever when the tip comes close to the surface, Van der Waals forces, dipole-dipole interactions, electrostatic forces, etc. cause the amplitude of the cantilever's oscillation to change (usually decrease) as the tip gets closer to the sample. This amplitude is used as the parameter that goes into the electronic servo that controls the height of the cantilever above the sample. The servo adjusts the height to maintain a set cantilever oscillation amplitude as the cantilever is scanned over the sample. A tapping AFM image is therefore produced by imaging the force of the intermittent contacts of the tip with the sample surface.

Although the peak forces applied during the contacting part of the oscillation can be much higher than typically used in contact mode, tapping mode generally lessens the damage done to the surface and the tip compared to the amount done in contact mode. This can be explained by the short duration of the applied force, and because the lateral forces between tip and sample are significantly lower in tapping mode over contact mode. Tapping mode imaging is gentle enough even for the visualization of supported lipid bilayers or adsorbed single polymer molecules (for instance, 0.4 nm thick chains of synthetic polyelectrolytes) under liquid medium. With proper scanning parameters, the conformation of single molecules can remain unchanged for hours, and even single molecular motors can be imaged while moving.

When operating in tapping mode, the phase of the cantilever›s oscillation with respect to the driving signal can be recorded as well. This signal channel contains information about the energy dissipated by the cantilever in each oscillation cycle. Samples that contain regions of varying stiffness or with different adhesion properties can give a contrast in this channel that is not visible in the topographic image. Extracting the sample›s material properties in a quantitative manner from phase images, however, is often not feasible.

Non-Contact Mode

In non-contact atomic force microscopy mode, the tip of the cantilever does not contact the sample surface. The cantilever is instead oscillated at either its resonant frequency (frequency modulation) or just above (amplitude modulation) where the amplitude of oscillation is typically a few nanometers (<10 nm) down to a few picometers. The van der Waals forces, which are strongest from 1 nm to 10 nm above the surface, or any other long-range force that extends above the surface acts to decrease the resonance frequency of the cantilever. This decrease in resonant frequency combined with the feedback loop system maintains a constant oscillation amplitude or frequency by adjusting the average tip-to-sample distance. Measuring the tip-to-sample distance at each (x,y) data point allows the scanning software to construct a topographic image of the sample surface.

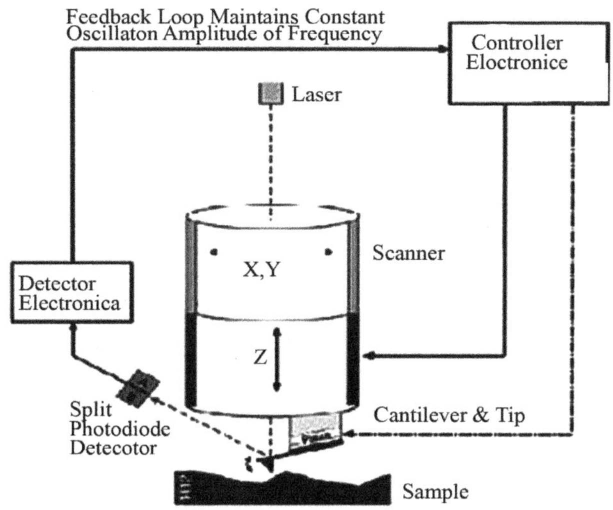

AFM – non-contact mode

Non-contact mode AFM does not suffer from tip or sample degradation effects that are sometimes observed after taking numerous scans with contact AFM. This makes non-contact AFM preferable to contact AFM for measuring soft samples, e.g. biological samples and organic thin film. In the case of rigid samples, contact and non-contact images may look the same. However, if a few monolayers of adsorbed fluid are lying on the surface of a rigid sample, the images may look quite different. An AFM operating in contact mode will penetrate the liquid layer to image the underlying surface, whereas in non-contact mode an AFM will oscillate above the adsorbed fluid layer to image both the liquid and surface.

Schemes for dynamic mode operation include frequency modulation where a phase-locked loop is used to track the cantilever's resonance frequency and the more common amplitude modulation with a servo loop in place to keep the cantilever excitation to a defined amplitude. In frequency modulation, changes in the oscillation frequency provide information about tip-sample interactions. Frequency can be measured with very high sensitivity and thus the frequency modulation mode allows for the use of very stiff cantilevers. Stiff cantilevers provide stability very close to the surface and, as a result, this technique was the first AFM technique to provide true atomic resolution in ultra-high vacuum conditions. In amplitude modulation, changes in the oscillation amplitude or phase provide the feedback signal for imaging. In amplitude modulation, changes in the phase of oscillation can be used to discriminate between different types of materials on the surface. Amplitude modulation can be operated either in the non-contact or in the intermittent contact regime. In dynamic contact mode, the cantilever is oscillated such that the separation distance between the cantilever tip and the sample surface is modulated.

Amplitude modulation has also been used in the non-contact regime to image with atomic resolution by using very stiff cantilevers and small amplitudes in an ultra-high vacuum environment.

Topographic Image

Image formation is a plotting method that produces a color mapping through changing the x-y position of the tip while scanning and recording the measured variable, i.e. the intensity of control signal, to each x-y coordinate. The color mapping shows the measured value corresponding to each coordinate. The image expresses the intensity of a value as a hue. Usually, the correspondence between the intensity of a value and a hue is shown as a color scale in the explanatory notes accompanying the image.

What is the Topographic Image of Atomic-Force Microscope?

Operation mode of Image forming of the AFM are generally classified into two groups from the viewpoint whether it uses z-Feedback loop (not shown) to maintain the tip-sample distance to keep signal intensity exported by the detector. The first one (using z-Feedback loop), said to be "constant **XX** mode" (**XX** is something which kept by z-Feedback loop).

Topographic Image Formation Mode is based on abovementioned "constant **XX** mode", z-Feedback loop controls the relative distance between the probe and the sample through outputting control signals to keep constant one of frequency, vibration and phase which typically corresponds to the motion of cantilever (for instance, voltage is applied to the Z-piezoelectric element and it moves the sample up and down towards the Z direction.

Details will be explained in the case that especially "constant df mode"(FM-AFM) among AFM as an instance in next section.

Topographic Image of FM-AFM

When the distance between the probe and the sample is brought to the range where atomic force may be detected, while a cantilever is excited in its natural eigen frequency (f_0), a phenomenon that the resonance frequency (f) of the cantilever shifts from the original resonance frequency (natural eigen frequency) of the cantilever. In other words, in the range where atomic force may be detected, the frequency shift ($df=f-f_0$) will be observed. So, when the distance between the probe and the sample is in the non-contact region, the frequency shift increases in negative direction as the distance between the probe and the sample gets smaller.

When the sample has concavity and convexity, the distance between the tip-apex and the sample varies in accordance with the concavity and convexity accompanied with a scan of the sample along x-y direction (without height regulation in z-direction) . As a result, the frequency shift arises. The image in which the values of the frequency obtained by a raster scan along the x-y direction of the sample surface are plotted against the x-y coordination of each measurement point is called a constant-height image.

On the other hand, the df may be kept constant by moving the probe upward and downward in z-direction using a negative feedback (by using z-feedback loop) while the raster scan of the sample surface along the x-y direction . The image in which the amounts of the negative feedback (the moving distance of the probe upward and downward in z-direction) are plotted against the x-y coordination of each measurement point is a topographic image. In other words, the topographic image is a trace of the tip of the probe regulated so that the df is constant and it may also be considered to be a plot of a constant-height surface of the df.

Therefore, the topographic image of the AFM is not the exact surface morphology itself, but actually the image influenced by the bond-order between the probe and the sample, however, the topographic image of the AFM is considered to reflect the geographical shape of the surface more than the topographic image of a scanning tunnel microscope.

Force Spectroscopy

Another major application of AFM (besides imaging) is force spectroscopy, the direct measurement of tip-sample interaction forces as a function of the gap between the tip and sample (the result of this measurement is called a force-distance curve). For this method, the AFM tip is extended towards and retracted from the surface as the deflection of the cantilever is monitored as a function of piezoelectric displacement. These measurements have been used to measure nanoscale contacts, atomic bonding, Van der Waals forces, and Casimir forces, dissolution forces in liquids and single molecule stretching and rupture forces. Furthermore, AFM was used to measure, in an aqueous environment, the dispersion force due to polymer adsorbed on the substrate. Forces of the order of a few piconewtons can now be routinely measured with a vertical distance resolution of better than 0.1 nanometers. Force spectroscopy can be performed with either static or dynamic modes. In dynamic modes, information about the cantilever vibration is monitored in addition to the static deflection.

Problems with the technique include no direct measurement of the tip-sample separation and the common need for low-stiffness cantilevers, which tend to 'snap' to the surface. These problems are not insurmountable. An AFM that directly measures the tip-sample separation has been developed. The snap-in can be reduced by measuring in liquids or by using stiffer cantilevers, but in the latter case a more sensitive deflection sensor is needed. By applying a small dither to the tip, the stiffness (force gradient) of the bond can be measured as well.

Biological Applications and Other

Force spectroscopy is used in biophysics to measure the mechanical properties. of living material (such as tissue or cells). Another application was to measure the interaction forces between from one hand a material stuck on the tip of the cantilever, and from another hand the surface of particles either free or occupied by the same material. From the adhesion force distribution curve, a mean value of the forces has been derived. It allowed to make a cartography of the surface of the particles, covered or not by the material.

Identification of Individual Surface Atoms

The AFM can be used to image and manipulate atoms and structures on a variety of surfaces. The atom at the apex of the tip "senses" individual atoms on the underlying surface when it forms incipient chemical bonds with each atom. Because these chemical interactions subtly alter the tip's vibration frequency, they can be detected and mapped.

This principle was used to distinguish between atoms of silicon, tin and lead on an alloy surface, by comparing these 'atomic fingerprints' to values obtained from large-scale density functional theory (DFT) simulations.

The trick is to first measure these forces precisely for each type of atom expected in the sample, and then to compare with forces given by DFT simulations. The team found that the tip interacted most strongly with silicon atoms, and interacted 24% and 41% less strongly with tin and lead atoms, respectively. Thus, each different type of atom can be identified in the matrix as the tip is moved across the surface.

Probe

An AFM probe has a sharp tip on the free-swinging end of a cantilever that is protruding from a holder. The dimensions of the cantilever are in the scale of micrometers. The radius of the tip is usually on the scale of a few nanometers to a few tens of nanometers. (Specialized probes exist with much larger end radii, for example probes for indentation of soft materials.) The cantilever holder, also called holder chip – often 1.6 mm by 3.4 mm in size – allows the operator to hold the AFM cantilever/probe assembly with tweezers and fit it into the corresponding holder clips on the scanning head of the atomic-force microscope.

This device is most commonly called an "AFM probe", but other names include "AFM tip" and "cantilever" (employing the name of a single part as the name of the whole device). An AFM probe is a particular type of SPM (scanning probe microscopy) probe.

AFM probes are manufactured with MEMS technology. Most AFM probes used are made from silicon (Si), but borosilicate glass and silicon nitride are also in use. AFM probes are considered consumables as they are often replaced when the tip apex becomes dull or contaminated or when the cantilever is broken. They can cost from a couple of tens of dollars up to hundreds of dollars per cantilever for the most specialized cantilever/probe combinations.

Just the tip is brought very close to the surface of the object under investigation, the cantilever is deflected by the interaction between the tip and the surface, which is what the AFM is designed to measure. A spatial map of the interaction can be made by measuring the deflection at many points on a 2D surface.

Several types of interaction can be detected. Depending on the interaction under investigation, the surface of the tip of the AFM probe needs to be modified with a coating. Among the coatings used are gold – for covalent bonding of biological molecules and the detection of their interaction with a surface, diamond for increased wear resistance and magnetic coatings for detecting the magnetic properties of the investigated surface. Another solution exists to achieve high resolution magnetic imaging : having the probe equip with a microSQUID. The AFM tips is fabricated using silicon micro machining and the precise positioning of the microSQUID loop is done by electron beam lithography.

The surface of the cantilevers can also be modified. These coatings are mostly applied in order to increase the reflectance of the cantilever and to improve the deflection signal.

AFM Cantilever-Deflection Measurement

Beam-Deflection Measurement

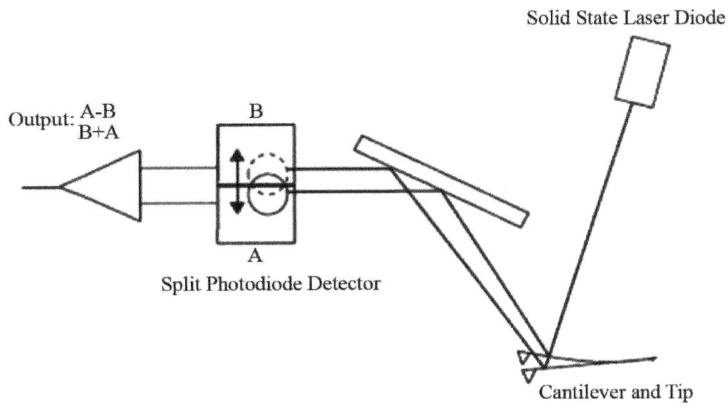

AFM beam-deflection detection

PSD consisting of two closely spaced photodiodes whose output signal is collected by a differential amplifier. Angular displacement of the cantilever results in one photodiode collecting more light than the other photodiode, producing an output signal (the difference between the photodiode signals normalized by their sum), which is proportional to the deflection of the cantilever. The sensitivity of the beam-deflection method is very high, a noise floor on the order of 10 fm $Hz^{-1/2}$ can be obtained routinely in a well-designed system. Although this method is sometimes called the 'optical lever' method, the signal is not amplified if the beam path is made longer. A longer beam path increases the motion of the reflected spot on the photodiodes, but also widens the spot by the same amount due to diffraction, so that the same amount of optical power is moved from one photodiode to the other. The 'optical leverage' (output signal of the detector divided by deflection of the cantilever) is inversely proportional to the numerical aperture of the beam focusing optics, as long as the focused laser spot is small enough to fall completely on the cantilever. It is also inversely proportional to the length of the cantilever. The most common method for cantilever-deflection measurements is the beam-deflection method. In this method, laser light from a solid-state diode is reflected off the back of the cantilever and collected by a position-sensitive detector.

The relative popularity of the beam-deflection method can be explained by its high sensitivity and simple operation, and by the fact that cantilevers do not require electrical contacts or other special treatments, and can therefore be fabricated relatively cheaply with sharp integrated tips.

Other Deflection-Measurement Methods

Many other methods for beam-deflection measurements exist.

- Piezoelectric detection – Cantilevers made from quartz (such as the qPlus configuration), or other piezoelectric materials can directly detect deflection as an electrical signal. Cantilever oscillations down to 10pm have been detected with this method.
- Laser Doppler vibrometry – A laser Doppler vibrometer can be used to produce very accurate deflection measurements for an oscillating cantilever (thus is only used in non-contact mode). This method is expensive and is only used by relatively few groups.
- STM — The first atomic microscope used an STM complete with its own feedback mechanism to measure deflection.This method is very difficult to implement, and is slow to react to deflection changes compared to modern methods.
- Optical interferometry – Optical interferometry can be used to measure cantilever deflection. Due to the nanometre scale deflections measured in AFM, the interferometer is running in the sub-fringe regime, thus, any drift in laser power or wavelength has strong effects on the measurement. For these reasons optical interferometer measurements must be done with great care (for example using index matching fluids between optical fibre junctions), with very stable lasers. For these reasons optical interferometry is rarely used.
- Capacitive detection – Metal coated cantilevers can form a capacitor with another contact located behind the cantilever.Deflection changes the distance between the contacts and can be measured as a change in capacitance.
- Piezoresistive detection – Cantilevers can be fabricated with piezoresistive elements that act as a strain gauge. Using a Wheatstone bridge, strain in the AFM cantilever due to deflection can be measured. This is not commonly used in vacuum applications, as the piezoresistive detection dissipates energy from the system affecting Q of the resonance.

Piezoelectric Scanners

AFM scanners are made from piezoelectric material, which expands and contracts proportionally to an applied voltage. Whether they elongate or contract depends upon the polarity of the voltage applied. Traditionally the tip or sample is mounted on a 'tripod' of three piezo crystals, with each responsible for scanning in the x,y and z directions. In 1986, the same year as the AFM was invented, a new piezoelectric scanner, the tube scanner, was developed for use in STM. Later tube scanners were incorporated into AFMs. The tube scanner can move the sample in the x, y, and z directions using a single tube piezo with a single interior contact and four external contacts. An advantage of the tube scanner compared to the original tripod design, is better vibrational isolation, resulting from the higher resonant frequency of the single element construction, in combination with a low resonant frequency isolation stage. A disadvantage is that the x-y motion can cause unwanted z motion resulting in distortion. Another popular design for AFM scanners is the flexure stage, which uses separate piezos for each axis, and couples them through a flexure mechanism.

Scanners are characterized by their sensitivity, which is the ratio of piezo movement to piezo voltage, i.e., by how much the piezo material extends or contracts per applied volt. Because of differences in material or size, the sensitivity varies from scanner to scanner. Sensitivity varies non-linearly with respect to scan size. Piezo scanners exhibit more sensitivity at the end than at the beginning of a scan. This causes the forward and reverse scans to behave differently and display hysteresis between the two scan directions. This can be corrected by applying a non-linear voltage to the piezo electrodes to cause linear scanner movement and calibrating the scanner accordingly. One disadvantage of this approach is that it requires re-calibration because the precise non-linear voltage needed to correct non-linear movement will change as the piezo ages (see below). This problem can be circumvented by adding a linear sensor to the sample stage or piezo stage to detect the true movement of the piezo. Deviations from ideal movement can be detected by the sensor and corrections applied to the piezo drive signal to correct for non-linear piezo movement. This design is known as a 'closed loop' AFM. Non-sensored piezo AFMs are referred to as ‹open loop› AFMs.

The sensitivity of piezoelectric materials decreases exponentially with time. This causes most of the change in sensitivity to occur in the initial stages of the scanner›s life. Piezoelectric scanners are run for approximately 48 hours before they are shipped from the factory so that they are past the point where they may have large changes in sensitivity. As the scanner ages, the sensitivity will change less with time and the scanner would seldom require recalibration, though various manufacturer manuals recommend monthly to semi-monthly calibration of open loop AFMs.

Advantages and Disadvantages

The first atomic-force microscope

Just like any other tool, an AFM's usefulness has limitations. When determining whether or not analyzing a sample with an AFM is appropriate, there are various advantages and disadvantages that must be considered.

Advantages

AFM has several advantages over the scanning electron microscope (SEM). Unlike the electron microscope, which provides a two-dimensional projection or a two-dimensional image of a sample, the AFM provides a three-dimensional surface profile. In addition, samples viewed by AFM do not require any special treatments (such as metal/carbon coatings) that would irreversibly change or damage the sample, and does not typically suffer from charging artifacts in the final image. While an electron microscope needs an expensive vacuum environment for proper operation, most AFM modes can work perfectly well in ambient air or even a liquid environment. This makes it possible to study biological macromolecules and even living organisms. In principle, AFM can provide higher resolution than SEM. It has been shown to give true atomic resolution in ultra-high vacuum (UHV) and, more recently, in liquid environments. High resolution AFM is comparable in resolution to scanning tunneling microscopy and transmission electron microscopy. AFM can also be combined with a variety of optical microscopy techniques such as fluorescent microscopy, further expanding its applicability. Combined AFM-optical instruments have been applied primarily in the biological sciences but have also found a niche in some materials applications, especially those involving photovoltaics research.

Disadvantages

A disadvantage of AFM compared with the scanning electron microscope (SEM) is the single scan image size. In one pass, the SEM can image an area on the order of square millimeters with a depth of field on the order of millimeters, whereas the AFM can only image a maximum scanning area of about 150×150 micrometers and a maximum height on the order of 10-20 micrometers. One method of improving the scanned area size for AFM is by using parallel probes in a fashion similar to that of millipede data storage.

The scanning speed of an AFM is also a limitation. Traditionally, an AFM cannot scan images as fast as an SEM, requiring several minutes for a typical scan, while an SEM is capable of scanning at near real-time, although at relatively low quality. The relatively slow rate of scanning during AFM imaging often leads to thermal drift in the image making the AFM less suited for measuring accurate distances between topographical features on the image. However, several fast-acting designs were suggested to increase microscope scanning productivity including what is being termed videoAFM (reasonable quality images are being obtained with videoAFM at video rate: faster than the average SEM). To eliminate image distortions induced by thermal drift, several methods have been introduced.

AFM images can also be affected by nonlinearity, hysteresis, and creep of the piezoelectric material and cross-talk between the x, y, z axes that may require software enhancement and filtering. Such filtering could "flatten" out real topographical features. However, newer AFMs utilize real-time correction software (for example, feature-oriented scanning) or closed-loop scanners, which practically eliminate these problems. Some AFMs also use separated orthogonal scanners (as opposed to a single tube), which also serve to eliminate part of the cross-talk problems.

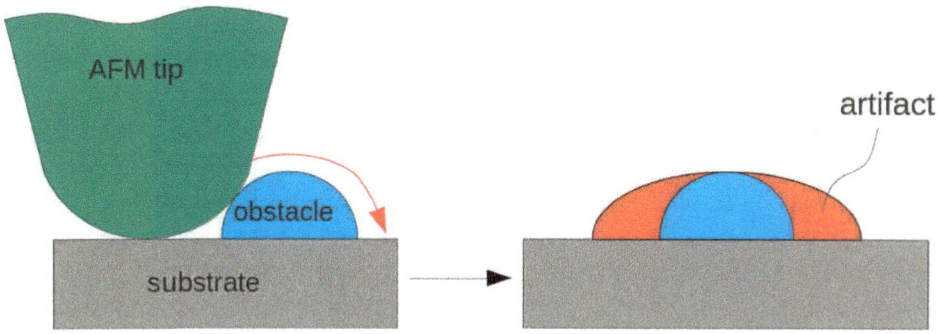

Showing an AFM artifact arising from a tip with a high radius of curvature with respect to the feature that is to be visualized.

As with any other imaging technique, there is the possibility of image artifacts, which could be induced by an unsuitable tip, a poor operating environment, or even by the sample itself, as depicted on the right. These image artifacts are unavoidable; however, their occurrence and effect on results can be reduced through various methods. Artifacts resulting from a too-coarse tip can be caused for example by inappropriate handling or de facto collisions with the sample by either scanning too fast or having an unreasonably rough surface, causing actual wearing of the tip.

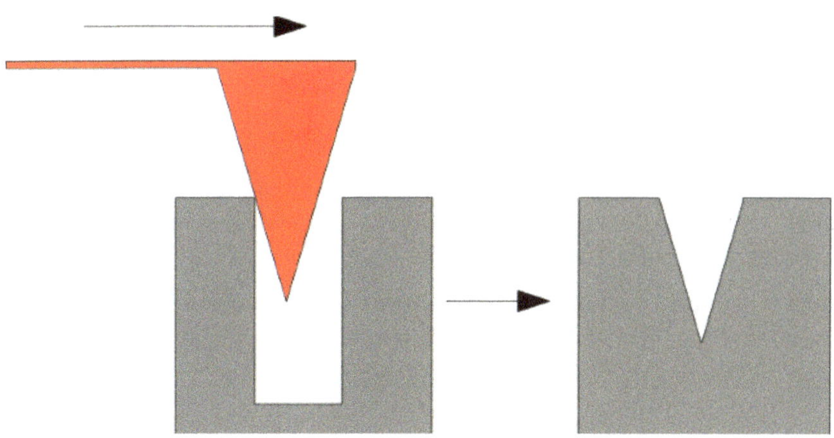

AFM artifact, steep sample topography

Due to the nature of AFM probes, they cannot normally measure steep walls or overhangs. Specially made cantilevers and AFMs can be used to modulate the probe sideways as well as up and down (as with dynamic contact and non-contact modes) to measure sidewalls, at the cost of more expensive cantilevers, lower lateral resolution and additional artifacts.

Other Applications in Various Fields of Study

The latest efforts in integrating nanotechnology and biological research have been successful and show much promise for the future. Since nanoparticles are a potential vehicle of drug delivery, the biological responses of cells to these nanoparticles are continuously being explored to optimize their efficacy and how their design could be improved. Pyrgiotakis et al. were able to study the interaction between CeO2 and Fe2O3 engineered nanoparticles and cells by attaching the engineered nanoparticles to the AFM tip. Beyond the interactions with external synthetic materials, cells have been imaged with X-ray crystallography and there has been much curiosity about their behavior in vivo. Studies have taken advantage of AFM to obtain further information on the behavior of live cells in biological media. Real-time atomic force spectroscopy (or nanoscopy) and dynamic atomic force spectroscopy have been used to study live cells and membrane proteins and their dynamic behavior at high resolution, on the nanoscale. Imaging and obtaining information on the topography and the properties of the cells has also given insight into chemical processes and mechanisms that occur through cell-cell interaction and interactions with other signaling molecules (ex. ligands). Evans and Calderwood used single cell force microscopy to study cell adhesion forces, bond kinetics/dynamic bond strength and its role in chemical processes such as cell signaling. Scheuring, Lévy, and Rigaud reviewed studies in which AFM to explore the crystal structure of membrane proteins of photosynthetic bacteria. Alsteen et al. have used AFM-based nanoscopy to perform a real-time analysis of the interaction between live mycobacteria and antimycobacterial drugs (specifically isoniazid, ethionamide, ethambutol, and streptomycine), which serves as an example of the more in-depth analysis of pathogen-drug interactions that can be done through AFM.

SMALL-ANGLE SCATTERING

Small-angle scattering (SAS) is a scattering technique based on deflection of collimated radiation away from the straight trajectory after it interacts with structures that are much larger than the wavelength of the radiation. The deflection is small (0.1-10°) hence the name small-angle. SAS techniques can give information about the size, shape and orientation of structures in a sample.

Small-angle scattering (SAS) is a powerful technique for investigating large-scale structures from 10 Å up to thousands and even several tens of thousands of angstroms. The most important feature of the SAS method is its potential for analyzing the inner structure of disordered systems, and frequently the application of this method is a unique way to obtain direct structural information on systems with random arrangement of density inhomogeneities in such large-scales.

Currently, the SAS technique, with its well-developed experimental and theoretical procedures and wide range of studied objects, is a self-contained branch of the structural analysis of matter. Reflecting these situations, the international meeting on the small-angle scattering studies have been held in every three years. SAS can refer to:

- Small angle neutron scattering (SANS)
- Small angle X-ray scattering (SAXS)

Applications

Small-angle scattering is particularly useful because of the dramatic increase in forward scattering that occurs at phase transitions, known as critical opalescence, and because many materials, substances and biological systems possess interesting and complex features in their structure, which match the useful length scale ranges that these techniques probe. The technique provides valuable information over a wide variety of scientific and technological applications including chemical aggregation, defects in materials, surfactants, colloids, ferromagnetic correlations in magnetism, alloy segregation, polymers, proteins, biological membranes, viruses, ribosome and macromolecules. While analysis of the data can give information on size, shape, etc., without making any model assumptions a preliminary analysis of the data can only give information on the radius of gyration for a particle using Guinier's equation.

Theory

Continuum Description

SAXS patterns are typically represented as scattered intensity as a function of the magnitude of the scattering vector $q = 4\pi \sin(\theta)/\lambda$. Here 2θ is the angle between the incident X-ray beam and the detector measuring the scattered intensity, and λ is the wavelength of the X-rays. One interpretation of the scattering vector is that it is the resolution or yardstick with which the sample is observed. In the case of a two-phase sample, e.g. small particles in liquid suspension, the only contrast leading to scattering in the typical range of resolution of the SAXS is simply $\Delta\rho$, the difference in average electron density between the particle and the surrounding liquid, because variations in ρ due to the atomic structure only become visible at higher angles in the WAXS regime. This means that the total integrated intensity of the SAXS pattern (in 3D) is an invariant quantity proportional to the square $\Delta\rho^2$. In 1-dimensional projection, as usually recorded for an isotropic pattern this invariant quantity becomes $\int I(q)q^2\, dx$, where the integral runs from q=0 to wherever the SAXS pattern is assumed to end and the WAXS pattern starts. It is also assumed that the density does not vary in the liquid or inside the particles, i.e. there is binary contrast.

SANS is described in terms of a neutron scattering length density.

Porod's Law

At wave numbers that are relatively large on the scale of SAS, but still small when compared to wide-angle Bragg diffraction, local interface intercorrelations are probed, whereas correlations between opposite interface segments are averaged out. For smooth interfaces, one obtains Porod's law:

$$I(q) \sim Sq^{-4}$$

This allows the surface area S of the particles to be determined with SAS. This needs to be modified if the interface is rough on the scale q^{-1}. If the roughness can be described by a fractal dimension d between 2-3 then Porod›s law becomes:

$$I(q) \sim S' q^{-(6-d)}$$

Scattering from Particles

Small-angle scattering from particles can be used to determine the particle shape or their size distribution. A small-angle scattering pattern can be fitted with intensities calculated from different model shapes when the size distribution is known. If the shape is known, a size distribution may be fitted to the intensity. Typically one assumes the particles to be spherical in the latter case.

If the particles are dispersed in a solution and they are known to be monodisperse, all of the same size, then a typical strategy is to measure different concentrations of particles in the solution. From the obtained SAXS patterns one can extrapolate to the intensity pattern one would get for a single particle. This is a necessary procedure that eliminates the concentration effect, which is a small shoulder that appears in the intensity patterns due to the proximity of neighbouring particles. The average distance between particles is then roughly the distance 2π/q*, where q* is the position of the shoulder on the scattering vector range q. The shoulder thus comes from the structure of the solution and this contribution is called the structure factor. One can write for the small-angle X-ray scattering intensity:

$$I(q) = P(q)S(q),$$

where

$I(q)$ is the intensity as a function of the magnitude q of the scattering vector

$P(q)$ is the form factor

and $S(q)$ is the structure factor.

When the intensities from low concentrations of particles are extrapolated to infinite dilution, the structure factor is equal to 1 and no longer disturbs the determination of the particle shape from the form factor $P(q)$. One can then easily apply the Guinier approximation (also called Guinier law, after André Guinier), which applies only at the very beginning of the scattering curve, at small q-values. According to the Guinier approximation the intensity at small q depends on the radius of gyration of the particle.

An important part of the particle shape determination is usually the distance distribution function $p(r)$ which may be calculated from the intensity using a Fourier transform

$$p(r) = \frac{r^2}{2\pi^2} \int_0^\infty I(q) \frac{\sin qr}{qr} q^2 \, dq.$$

The distance distribution function $p(r)$ is related to the frequency of certain distances r within the particle. Therefore it goes to zero at the largest diameter of the particle. It starts from zero at

$r = 0$ due to the multiplication by r^2. The shape of the $p(r)$- function already tells something about the shape of the particle. If the function is very symmetric, the particle is also highly symmetric, like a sphere. The distance distribution function should not be confused with the size distribution.

The particle shape analysis is especially popular in biological small-angle X-ray scattering, where one determines the shapes of proteins and other natural colloidal polymers.

Small-angle scattering studies were initiated by André Guinier (1937). Subsequently, Peter Debye, Otto Kratky, Günther Porod, R. Hosemann and others developed the theoretical and experimental fundamentals of the method and they were established until around 1960. Later on, new progress in refining the method began in the 1970s and is continuing today.

Organisations

As a 'low resolution' diffraction technique, the worldwide interests of the small-angle scattering community are promoted and coordinated by the Commission on Small-Angle Scattering of the International Union of Crystallography. There are also a number of community-led networks and projects. One such network, CanSAS - the acronym stands for Collective Action for Nomadic Small-Angle Scatterers, emphasising the global nature of the technique, champions the development of instrumental calibration standards and data file formats. It has also created a Web Portal to a variety of SAS-related resources.

SMALL-ANGLE X-RAY SCATTERING

Small-angle X-ray scattering (SAXS) is a small-angle scattering (SAS) technique where the elastic scattering of X-rays (wavelength 0.1 ... 0.2 nm) by a sample which has inhomogeneities in the nm-range, is recorded at very low angles (typically 0.1 - 10°). This angular range contains information about the shape and size of macromolecules, characteristic distances of partially ordered materials, pore sizes, and other data. SAXS is capable of delivering structural information of macromolecules between 5 and 25 nm, of repeat distances in partially ordered systems of up to 150 nm.USAXS (ultra-small angle X-ray scattering) can resolve even larger dimensions.

SAXS and USAXS belong to a family of X-ray scattering techniques that are used in the characterization of materials. In the case of biological macromolecules such as proteins, the advantage of SAXS over crystallography is that a crystalline sample is not needed. Nuclear magnetic resonance spectroscopy methods encounter problems with macromolecules of higher molecular mass (> 30-40 kDa). However, owing to the random orientation of dissolved or partially ordered molecules, the spatial averaging leads to a loss of information in SAXS compared to crystallography.

Applications

SAXS is used for the determination of the microscale or nanoscale structure of particle systems in terms of such parameters as averaged particle sizes, shapes, distribution, and surface-to-volume ratio. The materials can be solid or liquid and they can contain solid, liquid or gaseous domains (so-called particles) of the same or another material in any combination. Not only particles, but also the structure of ordered systems like lamellae, and fractal-like materials can be studied. The method is accurate, non-destructive and usually requires only a minimum of sample preparation. Applications are very broad and include colloids of all types, metals, cement, oil, polymers, plastics, proteins, foods and pharmaceuticals and can be found in research as well as in quality control. The X-ray source can be a laboratory source or synchrotron light which provides a higher X-ray flux.

SAXS Instruments

In an SAXS instrument a monochromatic beam of X-rays is brought to a sample from which some of the X-rays scatter, while most simply go through the sample without interacting with it. The scattered X-rays form a scattering pattern which is then detected at a detector which is typically a 2-dimensional flat X-ray detector situated behind the sample perpendicular to the direction of the primary beam that initially hit the sample. The scattering pattern contains the information on the structure of the sample. The major problem that must be overcome in SAXS instrumentation is the separation of the weak scattered intensity from the strong main beam. The smaller the desired angle, the more difficult this becomes. The problem is comparable to one encountered when trying to observe a weakly radiant object close to the sun, like the sun's corona. Only if the moon blocks out the main light source does the corona become visible. Likewise, in SAXS the non-scattered beam that merely travels through the sample must be blocked, without blocking the closely adjacent scattered radiation. Most available X-ray sources produce divergent beams and this compounds the problem. In principle the problem could be overcome by focusing the beam, but this is not easy when dealing with X-rays and was previously not done except on synchrotrons where large bent mirrors can be used. This is why most laboratory small angle devices rely on collimation instead. Laboratory SAXS instruments can be divided into two main groups: point-collimation and line-collimation instruments:

- **Point-collimation instruments** have pinholes that shape the X-ray beam to a small circular or elliptical spot that illuminates the sample. Thus the scattering is centro-symmetrically distributed around the primary X-ray beam and the scattering pattern in the detection plane consists of circles around the primary beam. Owing to the small illuminated sample volume and the wastefulness of the collimation process — only those photons are allowed to pass that happen to fly in the right direction — the scattered intensity is small and therefore the measurement time is in the order of hours or days in case of very weak scatterers. If focusing optics like bent mirrors or bent monochromator crystals or collimating and monochromating optics like multilayers are used, measurement time can be greatly reduced. Point-collimation allows the orientation of non-isotropic systems (fibres, sheared liquids) to be determined.
- **Line-collimation instruments** restrict the beam only in one dimension (rather than two as for point collimation) so that the beam cross-section is a long but narrow line. The illuminated sample volume is much larger compared to point-

collimation and the scattered intensity at the same flux density is proportionally larger. Thus measuring times with line-collimation SAXS instruments are much shorter compared to point-collimation and are in the range of minutes. A disadvantage is that the recorded pattern is essentially an integrated superposition (a self-convolution) of many adjacent pinhole patterns. The resulting smearing can be easily removed using model-free algorithms or deconvolution methods based on Fourier transformation, but only if the system is isotropic. Line collimation is of great benefit for any isotropic nanostructured materials, e.g. proteins, surfactants, particle dispersion and emulsions.

REFERENCES

1. Erni, Rolf; Rossell, MD; Kisielowski, C; Dahmen, U (2009). "Atomic-Resolution Imaging with a Sub-50-pm Electron Probe". Physical Review Letters. 102 (9): 096101. Bibcode:2009PhRvL.102i6101E. doi:10.1103/PhysRevLett.102.096101. PMID 19392535.
2. Mathys, Daniel, Zentrum für Mikroskopie, University of Basel: Die Entwicklung der Elektronenmikroskopie vom Bild über die Analyse zum Nanolabor, p. 8
3. Dannen, Gene (1998) Leo Szilard the Inventor: A Slideshow (1998, Budapest, conference talk). dannen.com
4. Ruska, Ernst (1986). "Ernst Ruska Autobiography". Nobel Foundation. Retrieved 2010-01-31.
5. Rudenberg, H Gunther; Rudenberg, Paul G (2010). "Chapter 6 – Origin and Background of the Invention of the Electron Microscope: Commentary and Expanded Notes on Memoir of Reinhold Rüdenberg". Advances in Imaging and Electron Physics. 160. Elsevier. doi:10.1016/S1076-5670(10)60006-7. ISBN 978-0-12-381017-5.
6. Kruger DH; Schneck P; Gelderblom HR (May 2000). "Helmut Ruska the visualisation of viruses". Lancet. 355 (9216): 1713–7. doi:10.1016/S0140-6736(00)02250-9. PMID 10905259.
7. von Ardenne, M; Beischer, D (1940). "Untersuchung von metalloxyd-rauchen mit dem universal-elektronenmikroskop". Zeitschrift Electrochemie (in German). 46: 270–277. doi:10.1002/bbpc.19400460406 (inactive 2016-09-13).
8. http://authors.library.caltech.edu/5456/1/hrst.mit.edu/hrs/materials/public/ElectronMicroscope/EM_HistOverview.htm
9. "James Hillier". Inventor of the Week: Archive. 2003-05-01. Retrieved 2010-01-31.
10. "The Scale of Things". Office of Basic Energy Sciences, U.S. Department of Energy. 2006-05-26. Retrieved 2010-01-31.
11. O'Keefe MA; Allard LF. "Sub-Ångstrom Electron Microscopy for Sub-Ångstrom Nano-Metrology" (pdf). Information Bridge: DOE Scientific and Technical Information – Sponsored by OSTI. Retrieved 2010-01-31.
12. Burgess, Jeremy (1987). Under the Microscope: A Hidden World Revealed.

CUP Archive. p. 11. ISBN 0-521-39940-8.

13. "Introduction to Electron Microscopy" (PDF). FEI Company. p. 15. Retrieved 12 December 2012.
14. Antonovsky, A. (1984). "The application of colour to sem imaging for increased definition". Micron and Microscopica Acta. 15 (2): 77–84. doi:10.1016/0739-6260(84)90005-4.
15. Danilatos, G.D. (1986). "Colour micrographs for backscattered electron signals in the SEM". Scanning. 9 (3): 8–18. doi:10.1111/j.1365-2818.1986.tb04287.x.
16. Danilatos, G.D. (1986). "Environmental scanning electron microscopy in colour". J. Microscopy. 142: 317–325. doi:10.1002/sca.4950080104.
17. "SPLEEM". National Center for Electron Microscopy (NCEM). Retrieved 2010-01-31.
18. Luft, J.H. (1961). "Improvements in epoxy resin embedding methods". The Journal of biophysical and biochemical cytology. 9 (2). p. 409. PMC 2224998 . PMID 13764136.
19. Juniper, B.E.; Bradley, D.E. (1958). "The carbon replica technique in the study of the ultrastructure of leaf surfaces". Journal of ultrastructure research. 2 (1): 16–27. doi:10.1016/s0022-5320(58)90045-5.
20. Reynolds, E. S. (1963). "The use of lead citrate at high pH as an electron-opaque stain in electron microscopy". Journal of Cell Biology. 17: 208–212. doi:10.1083/jcb.17.1.208. PMC 2106263 . PMID 13986422.
21. Adrian, Marc; Dubochet, Jacques; Lepault, Jean; McDowall, Alasdair W. (1984). "Cryo-electron microscopy of viruses". Nature. 308 (5954): 32–36. Bibcode:1984Natur.308...32A. doi:10.1038/308032a0. PMID 6322001.
22. Sabanay, I.; Arad, T.; Weiner, S.; Geiger, B. (1991). "Study of vitrified, unstained frozen tissue sections by cryoimmunoelectron microscopy". Journal of Cell Science. 100 (1): 227–236. PMID 1795028.
23. Kasas, S.; Dumas, G.; Dietler, G.; Catsicas, S.; Adrian, M. (2003). "Vitrification of cryoelectron microscopy specimens revealed by high-speed photographic imaging". Journal of Microscopy. 211 (1): 48–53. doi:10.1046/j.1365-2818.2003.01193.x.
24. Boehme, L.; Bresin, M.; Botman, A.; Ranney, J.; Hastings, J.T. (2015). "Focused electron beam induced etching of copper in sulfuric acid solutions". Nanotechnology. 26 (49): Article ID 495301. doi:10.1088/0957-4484/26/49/495301.
25. Kacher, J.; Cui, B.; Robertson, I.M. (2015). "In situ and tomographic characterization of damage and dislocation processes in irradiated metallic alloys by transmission electron microscopy". Journal of Materials Research. 30 (9): 1202–1213. doi:10.1557/jmr.2015.14.
26. Rai, R.S.; Subramanian, S. (2009). "Role of transmission electron microscopy in the semiconductor industry for process development and failure analysis". Progress in crystal growth and characterization of materials. 55 (3-4): 63–97. doi:10.1016/j.pcrysgrow.2009.09.002.
27. Morris, G.J.; Goodrich, M.; Acton, E.; Fonseca, F. (2006). "The high viscosity

encountered during freezing in glycerol solutions: Effects on cryopreservation". Cryobiology. 52 (3): 323–334. doi:10.1016/j.cryobiol.2006.01.003. PMID 16499898.

28. von Appen, A.; Beck, M. (2016). "Structure determination of the nuclear pore complex with three-dimensional cryo electron microscopy". Journal of Molecular Biology. 428 (10A): 2001–2010. doi:10.1016/j.jmb.2016.01.004. PMID 26791760.

29. Florian, P.E.; Rouillé, Y.; Ruta, S.; Nichita, N.; Roseanu, A. (2016). "Recent advances in human viruses imaging studies". Journal of Basic Microbiology. 56 (6): 591–607. doi:10.1002/jobm.201500575. PMID 27059598.

30. Cushnie, T.P.; O'Driscoll, N.H.; Lamb, A.J. (2016). "Morphological and ultrastructural changes in bacterial cells as an indicator of antibacterial mechanism of action". Cellular and Molecular Life Sciences. 73 (23): 4471–4492. doi:10.1007/s00018-016-2302-2. PMID 27392605.

31. Li, M.-H.; Yang, Y.-Q.; Huang, B.; Luo, X.; Zhang, W.; Han, M.; Ru, J.-G. (2014). "Development of advanced electron tomography in materials science based on TEM and STEM". Transactions of Nonferrous Metals Society of China. 24 (10): 3031–3050. doi:10.1016/S1003-6326(14)63441-5.

32. Li, W.J.; Shao, L.Y.; Zhang, D.Z.; Ro, C.U.; Hu, M.; Bi, X.H.; Geng, H.; Matsuki, A.; Niu, H.Y.; Chen, J.M. (2016). "A review of single aerosol particle studies in the atmosphere of East Asia: morphology, mixing state, source, and heterogeneous reactions". Journal of Cleaner Production. 112 (2): 1330–1349. doi:10.1016/j.jclepro2015.04.050.

33. Sousa, R.G.; Esteves, T.; Rocha, S.; Figueiredo, F.; Quelhas, P.; Silva, L.M. (2015). "Automatic detection of immunogold particles from electron microscopy images". Image Analysis and Recognition. 9164: 377–384. doi:10.1007/978-3-319-20801-5_41.

34. Perkins, G.A. (2014). "The use of miniSOG in the localization of mitochondrial proteins". Methods in Enzymology. 547: 165–179. doi:10.1016/B978-0-12-801415-8.00010-2. PMID 25416358.

35. Chen, X.D.; Ren, L.Q.; Zheng, B.; Liu, H. (2013). "Physics and engineering aspects of cell and tissue imaging systems: microscopic devices and computer assisted diagnosis". Biophotonics in Pathology: Pathology at the Crossroads. 185: 1–22. doi:10.3233/978-1-61499-234-9-1. PMID 23542929.

36. Fagerland, J.A.; Wall, H.G.; Pandher, K.; LeRoy, B.E.; Gagne, G.D. (2012). "Ultrastructural analysis in preclinical safety evaluation". Toxicologic Pathology. 40 (2): 391–402. doi:10.1177/0192623311430239. PMID 22215513.

37. Heider, S.; Metzner, C. (2014). "Quantitative real-time single particle analysis of virions". Virology. 462-463: 199–206. doi:10.1016/j.virol.2014.06.005. PMID 24999044.

38. Tsekouras, G.; Mozer, A.J.; Wallace, G.G. (2008). "Enhanced performance of dye sensitized solar cells utilizing platinum electrodeposit counter electrodes". Journal of The Electrochemical Society. 155 (7): K124–K128. doi:10.1149/1.2919107.

39. Besenius, P.; Portale, G.; Bomans, P.H.H.; Janssen, H.M.; Palmans, A.R.A.;

Meijer, E.W. (2010). "Controlling the growth and shape of chiral supramolecular polymers in water". Proceedings of the National Academy of Sciences of the United States of America. 107 (42): 17888–17893. doi:10.1073/pnas.1009592107.

40. Furuya, K. (2008). "Nanofabrication by advanced electron microscopy using intense and focused beam". Science and Technology of Advanced Materials. 9 (1): Article 014110. doi:10.1088/1468-6996/9/1/014110.

41. Maloy, S.A.; Sommer, W.F.; James, M.R.; Romero, T.J.; Lopez, M.R.; Zimmermann, E.; Ledbetter, J.M. (2000). "The accelerator production of tritium materials test program". Nuclear Technology. 132 (1): 103–114. ISSN 0029-5450.

42. Ukraintsev, V.; Banke, B. (2012). "Review of reference metrology for nanotechnology: significance, challenges, and solutions". Journal of Micro/Nanolithography, MEMS, and MOEMS. 11 (1): Article ID 011010. doi:10.1117/1.JMM.11.1.011010.

43. Wilhelmi, O.; Roussel, L.; Faber, P.; Reyntjens, S.; Daniel, G. (2010). "Focussed ion beam fabrication of large and complex nanopatterns". Journal of Experimental Nanoscience. 5 (3): 244–250. doi:10.1080/17458080903487448.

44. Vogt, E.T.C.; Whiting, G.T.; Chowdhury, A.D.; Weckhuysen, B.M. (2015). "Zeolites and zeotypes for oil and gas conversion". Advances in Catalysis. 58: 143–314. doi:10.1016/bs.acat.2015.10.001.

45. Lai, S.E.; Hong, Y.J.; Chen, Y.T.; Kang, Y.T.; Chang, P.; Yew, T.R. (2015). "Direct-writing of Cu nano-patterns with an electron beam". Microscopy and Microanalysis. 21 (6): 1639–1643. doi:10.1017/S1431927615015111.

46. Sicignano, A.; Di Monaco, R.; Masi, P.; Cavella, S. (2015). "From raw material to dish: pasta quality step by step". Journal of the Science of Food and Agriculture. 95 (13): 2579–2587. doi:10.1002/jsfa.7176. PMID 25783568.

47. Brożek-Mucha, Z. (2014). "Scanning electron microscopy and X-ray microanalysis for chemical and morphological characterisation of the inorganic component of gunshot residue: selected problems". Biomed Research International. 2014: Article ID 428038. doi:10.1155/2014/428038. PMID 25025050.

48. Carbonell-Verdu, A.; Garcia-Sanoguera, D.; Jorda-Vilaplana, A.; Sanchez-Nacher, L.; Balart, R. (2016). "A new biobased plasticizer for poly(vinyl chloride) based on epoxidized cottonseed oil". Journal of Applied Polymer Science. 33 (27): Article ID 43642. doi:10.1002/app.43642.

49. Ding, J.; Zhang, Z.M.; Wang, J.Q.; Han, E.H.; Tang, W.B.; Zhang, M.L.; Sun, Z.Y. (2015). "Micro-characterization of dissimilar metal weld joint for connecting the pipe-nozzle to the safe-end in generation III nuclear power plant". Acta Metallurgica Sinica. 51 (4): 425–439. doi:10.11900/0412.1961.2014.00299.

50. Tsikouras, B.; Pe-Piper, G.; Piper, D.J.W.; Schaffer, M. (2011). "Varietal heavy mineral analysis of sediment provenance, Lower Cretaceous Scotian Basin, eastern Canada". Sedimentary Geology. 237 (3-4): 150–165. doi:10.1016/j.sedgeo.2011.02.011.

51. Li, X.; Jiang, C.; Pan, L.L.; Zhang, H.Y.; Hu, L.; Li, T.X.; Yang, X.H. (2015). "Effects of preparing techniques and aging on dissolution behavior of the solid

dispersions of NF/Soluplus/Kollidon SR: identification and classification by a combined analysis by FT-IR spectroscopy and computational approaches". Drug Development and Industrial Pharmacy. 41 (1): 2–1. doi:10.3109/03639045.2014.938080. PMID 25026247.

52. Kepler J (1611). Strena seu de Nive Sexangula. Frankfurt: G. Tampach. ISBN 3-321-00021-0.
53. Steno N (1669). De solido intra solidum naturaliter contento dissertationis prodromus. Florentiae.
54. Hessel JFC (1831). Kristallometrie oder Kristallonomie und Kristallographie. Leipzig.
55. Bravais A (1850). "Mémoire sur les systèmes formés par des points distribués regulièrement sur un plan ou dans l'espace". Journal de l'Ecole Polytechnique. 19: 1.
56. Shafranovskii I I & Belov N V (1962). Paul Ewald, ed. "E. S. Fedorov" (PDF). 50 Years of X-Ray Diffraction. Springer: 351. ISBN 90-277-9029-9.
57. Schönflies A (1891). Kristallsysteme und Kristallstruktur. Leipzig.
58. Barlow W (1883). "Probable nature of the internal symmetry of crystals". Nature. 29 (738): 186. Bibcode:1883Natur..29..186B. doi:10.1038/029186a0.See also Barlow, William (1883). "Probable Nature of the Internal Symmetry of Crystals". Nature. 29 (739): 205. Bibcode:1883Natur..29..205B. doi:10.1038/029205a0. Sohncke, L. (1884). "Probable Nature of the Internal Symmetry of Crystals". Nature. 29 (747): 383. Bibcode:1884Natur..29..383S. doi:10.1038/029383a0. Barlow, WM. (1884). "Probable Nature of the Internal Symmetry of Crystals". Nature. 29 (748): 404. Bibcode:1884Natur..29..404B. doi:10.1038/029404b0.
59. Einstein A (1905). "Über einen die Erzeugung und Verwandlung des Lichtes betreffenden heuristischen Gesichtspunkt" [A Heuristic Model of the Creation and Transformation of Light]. Annalen der Physik (in German). 17 (6): 132. Bibcode:1905AnP...322..132E. doi:10.1002/andp.19053220607.. An English translation is available from Wikisource.
60. Compare: Einstein A (1909). "Über die Entwicklung unserer Anschauungen über das Wesen und die Konstitution der Strahlung" [The Development of Our Views on the Composition and Essence of Radiation)]. Physikalische Zeitschrift (in German). 10: 817.. An English translation is available from Wikisource.
61. Pais A (1982). Subtle is the Lord: The Science and the Life of Albert Einstein. Oxford University Press. ISBN 0-19-853907-X.
62. Compton A (1923). "A Quantum Theory of the Scattering of X-rays by Light Elements". Phys. Rev. 21 (5): 483. Bibcode:1923PhRv...21..483C. doi:10.1103/PhysRev.21.483.
63. Bragg WH (1907). "The nature of Röntgen rays". Transactions of the Royal Society of Science of Australia. 31: 94.
64. Bragg WH (1908). "The nature of γ- and X-rays". Nature. 77 (1995): 270. Bibcode:1908Natur..77..270B. doi:10.1038/077270a0. See also Bragg, W.

H. (1908). "The Nature of the γ and X-Rays". Nature. 78 (2021): 271. Bibcode:1908Natur..78..271B. doi:10.1038/078271a0.Bragg, W. H. (1908). "The Nature of the γ and X-Rays". Nature. 78 (2022): 293. Bibcode:1908Natur..78..293B. doi:10.1038/078293d0. Bragg, W. H. (1908). "The Nature of X-Rays". Nature. 78 (2035): 665. Bibcode:1908Natur..78R.665B. doi:10.1038/078665b0.

65. Bragg WH (1910). "The consequences of the corpuscular hypothesis of the γ- and X-rays, and the range of β-rays". Phil. Mag. 20 (117): 385. doi:10.1080/14786441008636917.
66. Bragg WH (1912). "On the direct or indirect nature of the ionization by X-rays". Phil. Mag. 23 (136): 647. doi:10.1080/14786440408637253.
67. Friedrich W; Knipping P; von Laue M (1912). "Interferenz-Erscheinungen bei Röntgenstrahlen". Sitzungsberichte der Mathematisch-Physikalischen Classe der Königlich-Bayerischen Akademie der Wissenschaften zu München. 1912: 303.
68. von Laue M (1914). "Concerning the detection of x-ray interferences" (PDF). Nobel Lectures, Physics. 1901–1921. Retrieved 2009-02-18.
69. Dana ES; Ford WE (1932). A Textbook of Mineralogy (fourth ed.). New York: John Wiley & Sons. p. 28.
70. Andre Guinier (1952). X-ray Crystallographic Technology. London: Hilger and Watts LTD. p. 271.
71. Bragg WL (1912). "The Specular Reflexion of X-rays". Nature. 90 (2250): 410. Bibcode:1912Natur..90..410B. doi:10.1038/090410b0.
72. Bragg WL (1913). "The Diffraction of Short Electromagnetic Waves by a Crystal". Proceedings of the Cambridge Philosophical Society. 17: 43.
73. Bragg (1914). "Die Reflexion der Röntgenstrahlen". Jahrbuch der Radioaktivität und Elektronik. 11: 350.
74. Bragg (1913). "The Structure of Some Crystals as Indicated by their Diffraction of X-rays". Proc. R. Soc. Lond. A89 (610): 248–277. Bibcode:1913RSPSA..89..248B. doi:10.1098/rspa.1913.0083. JSTOR 93488.
75. Bragg WL; James RW; Bosanquet CH (1921). "The Intensity of Reflexion of X-rays by Rock-Salt". Phil. Mag. 41 (243): 309. doi:10.1080/14786442108636225.
76. Bragg WL; James RW; Bosanquet CH (1921). "The Intensity of Reflexion of X-rays by Rock-Salt. Part II". Phil. Mag. 42 (247): 1. doi:10.1080/14786442108633730.
77. Bragg WL; James RW; Bosanquet CH (1922). "The Distribution of Electrons around the Nucleus in the Sodium and Chlorine Atoms". Phil. Mag. 44 (261): 433. doi:10.1080/14786440908565188.
78. Bragg WH; Bragg WL (1913). "The structure of the diamond". Nature. 91 (2283): 557. Bibcode:1913Natur..91..557B. doi:10.1038/091557a0.
79. Bragg WH; Bragg WL (1913). "The structure of the diamond". Proc. R. Soc. Lond. A89 (610): 277. Bibcode:1913RSPSA..89..277B. doi:10.1098/rspa.1913.0084.
80. Bragg WL (1914). "The Crystalline Structure of Copper". Phil. Mag. 28 (165): 355. doi:10.1080/14786440908635219.

81. Bragg WL (1914). "The analysis of crystals by the X-ray spectrometer". Proc. R. Soc. Lond. A89 (613): 468. Bibcode:1914RSPSA..89..468B. doi:10.1098/rspa.1914.0015.

82. Bragg WH (1915). "The structure of the spinel group of crystals". Phil. Mag. 30 (176): 305. doi:10.1080/14786440808635400.

83. Nishikawa S (1915). "Structure of some crystals of spinel group". Proc. Tokyo Math. Phys. Soc. 8: 199.

84. Vegard L (1916). "Results of Crystal Analysis". Phil. Mag. 32 (187): 65. doi:10.1080/14786441608635544.

85. Aminoff G (1919). "Crystal Structure of Pyrochroite". Stockholm Geol. Fören. Förh. 41: 407.

86. Aminoff G (1921). "Über die Struktur des Magnesiumhydroxids". Z. Kristallogr. 56: 505.

87. Bragg WL (1920). "The crystalline structure of zinc oxide". Phil. Mag. 39 (234): 647. doi:10.1080/14786440608636079.

88. Debije P, Scherrer P (1916). "Interferenz an regellos orientierten Teilchen im Röntgenlicht I". Physikalische Zeitschrift. 17: 277.

89. Friedrich W (1913). "Eine neue Interferenzerscheinung bei Röntgenstrahlen". Physikalische Zeitschrift. 14: 317.

90. Hull AW (1917). "A New Method of X-ray Crystal Analysis". Phys. Rev. 10 (6): 661. Bibcode:1917PhRv...10..661H. doi:10.1103/PhysRev.10.661.

91. Bernal JD (1924). "The Structure of Graphite". Proc. R. Soc. Lond. A106 (740): 749–773. JSTOR 94336.

92. Hassel O; Mack H (1924). "Über die Kristallstruktur des Graphits". Zeitschrift für Physik. 25: 317. Bibcode:1924ZPhy...25..317H. doi:10.1007/BF01327534.

93. Hull AW (1917). "The Crystal Structure of Iron". Phys. Rev. 9: 84. Bibcode:1917PhRv....9...83.. doi:10.1103/PhysRev.9.83.

94. Hull AW (1917). "The Crystal Structure of Magnesium". PNAS. 3 (7): 470. Bibcode:1917PNAS....3..470H. doi:10.1073/pnas.3.7.470.

95. Black, Susan AW (2005). "Domesticating the Crystal: Sir Lawrence Bragg and the Aesthetics of "X-ray Analysis"". Configurations. 13 (2): 257. doi:10.1353/con.2007.0014.

96. "From Atoms To Patterns". Wellcome Collection. Archived from the original on September 7, 2013. Retrieved 17 October 2013.

97. Wyckoff RWG; Posnjak E (1921). "The Crystal Structure of Ammonium Chloroplatinate". J. Amer. Chem. Soc. 43 (11): 2292. doi:10.1021/ja01444a002.

98. Bragg WH (1921). "The structure of organic crystals". Proc. R. Soc. Lond. 34: 33. Bibcode:1921PPSL...34...33B. doi:10.1088/1478-7814/34/1/306.

99. Lonsdale K (1928). "The structure of the benzene ring". Nature. 122 (3082): 810. Bibcode:1928Natur.122..810L. doi:10.1038/122810c0.

100. Pauling L. The Nature of the Chemical Bond (3rd ed.). Ithaca, NY: Cornell

University Press. ISBN 0-8014-0333-2.

101. Bragg WH (1922). "The crystalline structure of anthracene". Proc. R. Soc. Lond. 35: 167. Bibcode:1922PPSL...35..167B. doi:10.1088/1478-7814/35/1/320.

102. Powell HM; Ewens RVG (1939). "The crystal structure of iron enneacarbonyl". J. Chem. Soc.: 286. doi:10.1039/jr9390000286.

103. Bertrand JA, Cotton, Dollase (1963). "The Metal-Metal Bonded, Polynuclear Complex Anion in CsReCl4". J. Amer. Chem. Soc. 85 (9): 1349. doi:10.1021/ja00892a029.

104. Robinson WT; Fergusson JE; Penfold BR (1963). "Configuration of Anion in CsReCl4". Proceedings of the Chemical Society of London: 116.

105. Cotton FA, Curtis, Harris, Johnson, Lippard, Mague, Robinson, Wood (1964). "Mononuclear and Polynuclear Chemistry of Rhenium (III): Its Pronounced Homophilicity". Science. 145 (3638): 1305–7. Bibcode:1964Sci...145.1305C. doi:10.1126/science.145.3638.1305. PMID 17802015.

106. Cotton FA, Harris (1965). "The Crystal and Molecular Structure of Dipotassium Octachlorodirhenate(III) Dihydrate". Inorganic Chemistry. 4 (3): 330. doi:10.1021/ic50025a015.

107. Cotton FA (1965). "Metal-Metal Bonding in [Re2X8]2− Ions and Other Metal Atom Clusters". Inorganic Chemistry. 4 (3): 334. doi:10.1021/ic50025a016.

108. Eberhardt WH; Crawford W, Jr.; Lipscomb WN (1954). "The valence structure of the boron hydrides". J. Chem. Phys. 22 (6): 989. Bibcode:1954JChPh..22..989E. doi:10.1063/1.1740320.

109. Martin TW; Derewenda ZS (1999). "The name is Bond—H bond". Nature Structural Biology. 6 (5): 403–6. doi:10.1038/8195. PMID 10331860.

110. Dunitz JD; Orgel LE; Rich A (1956). "The crystal structure of ferrocene". Acta Crystallographica. 9 (4): 373. doi:10.1107/S0365110X56001091.

111. Seiler P; Dunitz JD (1979). "A new interpretation of the disordered crystal structure of ferrocene". Acta Crystallographica B. 35 (5): 1068. doi:10.1107/S0567740879005598.

112. Wunderlich JA; Mellor DP (1954). "A note on the crystal structure of Zeise's salt". Acta Crystallographica. 7: 130. doi:10.1107/S0365110X5400028X.

113. Jarvis JAJ; Kilbourn BT; Owston PG (1970). "A re-determination of the crystal and molecular structure of Zeise's salt, KPtCl3.C2H4.H2O. A correction". Acta Crystallographica B. 26 (6): 876. doi:10.1107/S056774087000328X.

114. Jarvis JAJ; Kilbourn BT; Owston PG (1971). "A re-determination of the crystal and molecular structure of Zeise's salt, KPtCl3.C2H4.H2O". Acta Crystallographica B. 27 (2): 366. doi:10.1107/S0567740871002231.

115. Love RA; Koetzle TF; Williams GJB; Andrews LC; Bau R (1975). "Neutron diffraction study of the structure of Zeise's salt, KPtCl3(C2H4).H2O". Inorganic Chemistry. 14 (11): 2653. doi:10.1021/ic50153a012.

116. Brown, Dwayne (October 30, 2012). "NASA Rover's First Soil Studies Help Fingerprint Martian Minerals". NASA. Retrieved October 31, 2012.

117. Westgren A; Phragmén G (1925). "X-ray Analysis of the Cu-Zn, Ag-Zn and Au-Zn Alloys". Phil. Mag. 50: 311.

118. Bradley AJ; Thewlis J (1926). "The structure of γ-Brass". Proc. R. Soc. Lond. 112 (762): 678. Bibcode:1926RSPSA.112..678B. doi:10.1098/rspa.1926.0134.

119. Hume-Rothery W (1926). "Researches on the Nature, Properties and Conditions of Formation of Intermetallic Compounds (with special Reference to certain Compounds of Tin)". Journal of the Institute of Metals. 35: 295.

120. Bradley AJ; Gregory CH (1927). "The Structure of certain Ternary Alloys". Nature. 120 (3027): 678. Bibcode:1927Natur.120..678.. doi:10.1038/120678a0.

121. Westgren A (1932). "Zur Chemie der Legierungen". Angewandte Chemie. 45 (2): 33. doi:10.1002/ange.19320450202.

122. Bernal JD (1935). "The Electron Theory of Metals". Annual Reports on the Progress of Chemistry. 32: 181.

123. Pauling L (1923). "The Crystal Structure of Magnesium Stannide". J. Amer. Chem. Soc. 45 (12): 2777. doi:10.1021/ja01665a001.

124. Pauling L (1929). "The Principles Determining the Structure of Complex Ionic Crystals". J. Amer. Chem. Soc. 51 (4): 1010. doi:10.1021/ja01379a006.

125. Dickinson RG; Raymond AL (1923). "The Crystal Structure of Hexamethylene-Tetramine". J. Amer. Chem. Soc. 45: 22. doi:10.1021/ja01654a003.

126. Müller A (1923). "The X-ray Investigation of Fatty Acids". Journal of the Chemical Society (London). 123: 2043. doi:10.1039/ct9232302043.

127. Saville WB; Shearer G (1925). "An X-ray Investigation of Saturated Aliphatic Ketones". Journal of the Chemical Society (London). 127: 591. doi:10.1039/ct9252700591.

128. Bragg WH (1925). "The Investigation of thin Films by Means of X-rays". Nature. 115 (2886): 266. Bibcode:1925Natur.115..266B. doi:10.1038/115266a0.

129. de Broglie M, Trillat JJ (1925). "Sur l'interprétation physique des spectres X d'acides gras". Comptes rendus hebdomadaires des séances de l'Académie des sciences. 180: 1485.

130. Trillat JJ (1926). "Rayons X et Composeés organiques à longe chaine. Recherches spectrographiques sue leurs structures et leurs orientations". Annales de Physique. 6: 5.

131. Caspari WA (1928). "Crystallography of the Aliphatic Dicarboxylic Acids". Journal of the Chemical Society (London). ?: 3235. doi:10.1039/jr9280003235.

132. Müller A (1928). "X-ray Investigation of Long Chain Compounds (n. Hydrocarbons)". Proc. R. Soc. Lond. 120 (785): 437. Bibcode:1928RSPSA.120..437M. doi:10.1098/rspa.1928.0158.

133. Piper SH (1929). "Some Examples of Information Obtainable from the long Spacings of Fatty Acids". Transactions of the Faraday Society. 25: 348. doi:10.1039/tf9292500348.

134. Müller A (1929). "The Connection between the Zig-Zag Structure of

the Hydrocarbon Chain and the Alternation in the Properties of Odd and Even Numbered Chain Compounds". Proc. R. Soc. Lond. 124 (794): 317. Bibcode:1929RSPSA.124..317M. doi:10.1098/rspa.1929.0117.

135. Robertson JM (1936). "An X-ray Study of the Phthalocyanines, Part II". Journal of the Chemical Society: 1195.
136. Crowfoot Hodgkin D (1935). "X-ray Single Crystal Photographs of Insulin". Nature. 135 (3415): 591. Bibcode:1935Natur.135..591C. doi:10.1038/135591a0.
137. Kendrew J. C., et al. (1958-03-08). "A Three-Dimensional Model of the Myoglobin Molecule Obtained by X-Ray Analysis". Nature. 181 (4610): 662–6. Bibcode:1958Natur.181..662K. doi:10.1038/181662a0. PMID 13517261.
138. "Table of entries in the PDB, arranged by experimental method".
139. "PDB Statistics". RCSB Protein Data Bank. Retrieved 2010-02-09.
140. Scapin G (2006). "Structural biology and drug discovery". Curr. Pharm. Des. 12 (17): 2087–97. doi:10.2174/138161206777585201. PMID 16796557.
141. Lundstrom K (2006). "Structural genomics for membrane proteins". Cell. Mol. Life Sci. 63 (22): 2597–607. doi:10.1007/s00018-006-6252-y. PMID 17013556.
142. Lundstrom K (2004). "Structural genomics on membrane proteins: mini review". Comb. Chem. High Throughput Screen. 7 (5): 431–9. doi:10.2174/1386207043328634. PMID 15320710.
143. Cryogenic (<20 K) helium cooling mitigates radiation damage to protein crystals" Acta Crystallographica Section D. 2007 63 (4) 486-492
144. J. Claydon, N. Greeves, S. Warren: Organic Chemistry 2nd edition page 45; Oxford University Press 2012
145. Greninger AB (1935). "A back-reflection Laue method for determining crystal orientation". Zeitschrift für Kristallographie. 91: 424.
146. An analogous diffraction pattern may be observed by shining a laser pointer on a compact disc or DVD; the periodic spacing of the CD tracks corresponds to the periodic arrangement of atoms in a crystal.
147. Miao, J., Charalambous, P., Kirz, J., & Sayre, D. (1999). "Extending the methodology of X-ray crystallography to allow imaging of micrometre-sized non-crystalline specimens."Nature, 400(6742), 342.
148. Harp, JM; Timm, DE; Bunick, GJ (1998). "Macromolecular crystal annealing: overcoming increased mosaicity associated with cryocrystallography". Acta Crystallographica D. 54 (Pt 4): 622–8. doi:10.1107/S0907444997019008. PMID 9761858.
149. Harp, JM; Hanson, BL; Timm, DE; Bunick, GJ (1999). "Macromolecular crystal annealing: evaluation of techniques and variables". Acta Crystallographica D. 55 (Pt 7): 1329–34. doi:10.1107/S0907444999005442. PMID 10393299.
150. Hanson, BL; Harp, JM; Bunick, GJ (2003). "The well-tempered protein crystal: annealing macromolecular crystals". Methods in enzymology. Methods in Enzymology. 368: 217–35. doi:10.1016/S0076-6879(03)68012-2. ISBN 978-0-12-182271-2. PMID 14674276.

151. Geerlof A, et al. (2006). "The impact of protein characterization in structural proteomics". Acta Crystallographica D. 62 (Pt 10): 1125–36. doi:10.1107/S0907444906030307. PMID 17001090.

152. Chernov AA (2003). "Protein crystals and their growth". J. Struct. Biol. 142 (1): 3–21. doi:10.1016/S1047-8477(03)00034-0. PMID 12718915.

153. Rupp B; Wang J (2004). "Predictive models for protein crystallization". Methods. 34 (3): 390–407. doi:10.1016/j.ymeth.2004.03.031. PMID 15325656.

154. Chayen NE (2005). "Methods for separating nucleation and growth in protein crystallization". Prog. Biophys. Mol. Biol. 88 (3): 329–37. doi:10.1016/j.pbiomolbio.2004.07.007. PMID 15652248.

155. Stock D; Perisic O; Lowe J (2005). "Robotic nanolitre protein crystallisation at the MRC Laboratory of Molecular Biology". Prog Biophys Mol Biol. 88 (3): 311–27. doi:10.1016/j.pbiomolbio.2004.07.009. PMID 15652247.

156. Jeruzalmi D (2006). "First analysis of macromolecular crystals: biochemistry and x-ray diffraction". Methods Mol. Biol. 364: 43–62. doi:10.1385/1-59745-266-1:43. ISBN 1-59745-266-1. PMID 17172760.

157. Helliwell JR (2005). "Protein crystal perfection and its application". Acta Crystallographica D. 61 (Pt 6): 793–8. doi:10.1107/S0907444905001368. PMID 15930642.

158. Garman, E. F.; Schneider, T. R. (1997). "Macromolecular Cryocrystallography". Journal of Applied Crystallography. 30 (3): 211. doi:10.1107/S0021889897002677.

159. Schlichting, I; Miao, J (2012). "Emerging opportunities in structural biology with X-ray free-electron lasers". Current Opinion in Structural Biology. 22 (5): 613–26. doi:10.1016/j.sbi.2012.07.015. PMC 3495068 . PMID 22922042.

160. Neutze, R; Wouts, R; Van Der Spoel, D; Weckert, E; Hajdu, J (2000). "Potential for biomolecular imaging with femtosecond X-ray pulses". Nature. 406 (6797): 752–7. Bibcode:2000Natur.406..752N. doi:10.1038/35021099. PMID 10963603.

161. Liu, W; Wacker, D; Gati, C; Han, G. W.; James, D; Wang, D; Nelson, G; Weierstall, U; Katritch, V; Barty, A; Zatsepin, N. A.; Li, D; Messerschmidt, M; Boutet, S; Williams, G. J.; Koglin, J. E.; Seibert, M. M.; Wang, C; Shah, S. T.; Basu, S; Fromme, R; Kupitz, C; Rendek, K. N.; Grotjohann, I; Fromme, P; Kirian, R. A.; Beyerlein, K. R.; White, T. A.; Chapman, H. N.; et al. (2013). "Serial femtosecond crystallography of G protein-coupled receptors". Science. 342 (6165): 1521–4. Bibcode:2013Sci...342.1521L. doi:10.1126/science.1244142. PMC 3902108 . PMID 24357322.

162. Ravelli RB; Garman EF (2006). "Radiation damage in macromolecular cryocrystallography". Curr. Opin. Struct. Biol. 16 (5): 624–9. doi:10.1016/j.sbi.2006.08.001. PMID 16938450.

163. Powell HR (1999). "The Rossmann Fourier autoindexing algorithm in MOSFLM". Acta Crystallographica D. 55 (Pt 10): 1690–5. doi:10.1107/S0907444999009506. PMID 10531518.

164. Hauptman H (1997). "Phasing methods for protein crystallography". Curr. Opin.

Struct. Biol. 7 (5): 672–80. doi:10.1016/S0959-440X(97)80077-2. PMID 9345626.

165. Usón I; Sheldrick GM (1999). "Advances in direct methods for protein crystallography". Curr. Opin. Struct. Biol. 9 (5): 643–8. doi:10.1016/S0959-440X(99)00020-2. PMID 10508770.
166. Taylor G (2003). "The phase problem". Acta Crystallographica D. 59 (11): 1881. doi:10.1107/S0907444903017815.
167. Ealick SE (2000). "Advances in multiple wavelength anomalous diffraction crystallography". Current Opinion in Chemical Biology. 4 (5): 495–9. doi:10.1016/S1367-5931(00)00122-8. PMID 11006535.
168. Parkin, G., "Bond-stretch isomerism in transition metal complexes: a reevaluation of crystallographic data", Chem. Rev. 1993, volume 93, 887-911. doi:10.1021/cr00019a003
169. Patterson AL (1935). "A Direct Method for the Determination of the Components of Interatomic Distances in Crystals". Zeitschrift für Kristallographie. 90: 517. doi:10.1524/zkri.1935.90.1.517.
170. "The Nobel Prize in Physics 1914". Nobel Foundation. Retrieved 2008-10-09.
171. "The Nobel Prize in Physics 1915". Nobel Foundation. Retrieved 2008-10-09.
172. "The Nobel Prize in Chemistry 1962". Nobelprize.org. Retrieved 2008-10-06.
173. "The Nobel Prize in Physiology or Medicine 1962". Nobel Foundation. Retrieved 2007-07-28.
174. "The Nobel Prize in Chemistry 1964". Nobelprize.org. Retrieved 2008-10-06.
175. "The Nobel Prize in Chemistry 1972". Nobelprize.org. Retrieved 2008-10-06.
176. "The Nobel Prize in Chemistry 1976". Nobelprize.org. Retrieved 2008-10-06.
177. "The Nobel Prize in Chemistry 1985". Nobelprize.org. Retrieved 2008-10-06.
178. "The Nobel Prize in Chemistry 1988". Nobelprize.org. Retrieved 2008-10-06.
179. "The Nobel Prize in Chemistry 1997". Nobelprize.org. Retrieved 2008-10-06.
180. "The Nobel Prize in Chemistry 2003". Nobelprize.org. Retrieved 2008-10-06.
181. "The Nobel Prize in Chemistry 2006". Nobelprize.org. Retrieved 2008-10-06.
182. "The Nobel Prize in Chemistry 2009". Nobelprize.org. Retrieved 2009-10-07.
183. "The Nobel Prize in Chemistry 2012". Nobelprize.org. Retrieved 2012-10-13.
184. "Background and Theory Page of Nuclear Magnetic Resonance Facility". Mark Wainwright Analytical Centre - University of Southern Wales Sydney. 9 December 2011. Retrieved 9 February 2014.
185. Shah, N; Sattar, A; Benanti, M; Hollander, S; Cheuck, L (January 2006). "Magnetic resonance spectroscopy as an imaging tool for cancer: a review of the literature.". The Journal of the American Osteopathic Association. 106 (1): 23–27. PMID 16428685.
186. Johnson Jr., C. S. (1999). "Diffusion ordered nuclear magnetic resonance spectroscopy: principles and applications". Progress in Nuclear Magnetic Resonance Spectroscopy. 34: 203–256. doi:10.1016/S0079-6565(99)00003-5.

187. Neufeld, R.; Stalke, D. (2015). "Accurate Molecular Weight Determination of Small Molecules via DOSY-NMR by Using External Calibration Curves with Normalized Diffusion Coefficients". Chem. Sci. 6: 3354–3364. doi:10.1039/C5SC00670H.

188. http://nmr.chem.wsu.edu/tutorials/basics/lock/

189. http://www2.chemistry.msu.edu/facilities/nmr/NMR_Artifacts.html

190. http://triton.iqfr.csic.es/guide/eNMR/eNMR1D/invrec.html

191. James Keeler. "Chapter 2: NMR and energy levels" (reprinted at University of Cambridge). Understanding NMR Spectroscopy. University of California, Irvine. Retrieved 2007-05-11.

192. Pople, J.A.; Bernstein, H. J.; Schneider, W. G. (1957). "The Analysis of Nuclear Magnetic Resonanace Spectra". Can J. Chem. 35: 65–81.

193. Aue, W. P. and Bartholdi, E. and Ernst, R. R., Two-dimensional spectroscopy. Application to nuclear magnetic resonance; The Journal of Chemical Physics, 64, 2229-2246 (1976)

194. Jeener, J., Jeener, Jean: Reminiscences about the Early Days of 2D NMR; John Wiley & Sons, Ltd: Encyclopedia of Magnetic Resonance (2007)

195. Martin, G.E; Zekter, A.S., Two-Dimensional NMR Methods for Establishing Molecular Connectivity; VCH Publishers, Inc: New York, 1988 (p.59)

196. A. Rahmani, C. Knight , and M. R. Morrow. Response to hydrostatic pressure of bicellar dispersions containing anionic lipid: Pressure-induced interdigitation. 2013, 29 (44), pp 13481–13490, DOI: 10.1021/la4035694

197. Fürtig, Boris; Richter, Christian; Wöhnert, Jens; Schwalbe, Harald (2003). "NMR Spectroscopy of RNA". ChemBioChem. 4 (10): 936–62. doi:10.1002/cbic.200300700. PMID 14523911.

198. Addess, Kenneth J.; Feigon, Juli (1996). "Introduction to 1H NMR Spectroscopy of DNA". In Hecht, Sidney M. Bioorganic Chemistry: Nucleic Acids. New York: Oxford University Press. ISBN 0-19-508467-5.

199. Wemmer, David (2000). "Chapter 5: Structure and Dynamics by NMR". In Bloomfield, Victor A.; Crothers, Donald M.; Tinoco, Ignacio. Nucleic acids: Structures, Properties, and Functions. Sausalito, California: University Science Books. ISBN 0-935702-49-0.

200. Patent US4724318 - Atomic-force microscope and method for imaging surfaces with atomic resolution

201. Binnig, G.; Quate, C. F.; Gerber, C. (1986). "Atomic-Force Microscope". Physical Review Letters. 56: 930–933. Bibcode:1986PhRvL..56..930B. doi:10.1103/physrevlett.56.930.

202. Lang, K.M.; D. A. Hite; R. W. Simmonds; R. McDermott; D. P. Pappas; John M. Martinis (2004). "Conducting atomic-force microscopy for nanoscale tunnel barrier characterization". Review of Scientific Instruments. 75 (8): 2726–2731. Bibcode:2004RScI...75.2726L. doi:10.1063/1.1777388.

203. Cappella, B; Dietler, G (1999). "Force-distance curves by atomic-force microscopy"

(PDF). Surface Science Reports. 34 (1–3): 1–104. Bibcode:1999SurSR..34....1C. doi:10.1016/S0167-5729(99)00003-5. Archived from the original (PDF) on 2012-12-03.

204. Binnig, G.; Quate, C. F.; Gerber, Ch. (1986). "Atomic-Force Microscope". Physical Review Letters. 56 (9): 930–933. Bibcode:1986PhRvL..56..930B. doi:10.1103/PhysRevLett.56.930. ISSN 0031-9007. PMID 10033323.

205. Roiter, Y; Minko, S (Nov 2005). "AFM single molecule experiments at the solid-liquid interface: in situ conformation of adsorbed flexible polyelectrolyte chains". Journal of the American Chemical Society. 127 (45): 15688–9. doi:10.1021/ja0558239. ISSN 0002-7863. PMID 16277495.

206. Zhong, Q; Inniss, D; Kjoller, K; Elings, V (1993). "Fractured polymer/silica fiber surface studied by tapping mode atomic-force microscopy". Surface Science Letters. 290: L688. Bibcode:1993SurSL.290L.688Z. doi:10.1016/0167-2584(93)90906-Y.

207. Geisse, Nicholas A. (July–August 2009). "AFM and Combined Optical Techniques". Materials Today. 12 (7–8): 40–45. doi:10.1016/S1369-7021(09)70201-9. Retrieved 4 November 2011.

208. Gross, L.; Mohn, F.; Moll, N.; Liljeroth, P.; Meyer, G. (27 August 2009). "The Chemical Structure of a Molecule Resolved by Atomic-Force Microscopy". Science. 325 (5944): 1110–1114. Bibcode:2009Sci...325.1110G. doi:10.1126/science.1176210. PMID 19713523.

209. Giessibl, Franz J. (2003). "Advances in atomic-force microscopy". Reviews of Modern Physics. 75 (3): 949–983. arXiv:cond-mat/0305119 . Bibcode:2003RvMP...75..949G. doi:10.1103/RevModPhys.75.949.

210. Hinterdorfer, P; Dufrêne, Yf (May 2006). "Detection and localization of single molecular recognition events using atomic-force microscopy". Nature Methods. 3 (5): 347–55. doi:10.1038/nmeth871. ISSN 1548-7091. PMID 16628204.

211. Ferrari, L.; Kaufmann, J.; Winnefeld, F.; Plank, J. (Jul 2010). "Interaction of cement model systems with superplasticizers investigated by atomic-force microscopy, zeta potential, and adsorption measurements". J Colloid Interface Sci. 347 (1): 15–24. doi:10.1016/j.jcis.2010.03.005. PMID 20356605.

212. Butt, H; Cappella, B; Kappl, M (2005). "Force measurements with the atomic-force microscope: Technique, interpretation and applications". Surface Science Reports. 59: 1–152. Bibcode:2005SurSR..59....1B. doi:10.1016/j.surfrep.2005.08.003.

213. Gavin M. King; Ashley R. Carter; Allison B. Churnside; Louisa S. Eberle & Thomas T. Perkins (2009). "Ultrastable Atomic-Force Microscopy: Atomic-Scale Stability and Registration in Ambient Conditions". Nano Letters. 9 (4): 1451–1456. Bibcode:2009NanoL...9.1451K. doi:10.1021/nl803298q. PMC 2953871 . PMID 19351191.

214. M. Hoffmann, Ahmet Oral; Ralph A. G, Peter (2001). "Direct measurement of interatomic-force gradients using an ultra-low-amplitude atomic-force microscope". Proceedings of the Royal Society A. 457 (2009): 1161–1174. Bibcode:2001RSPSA.457.1161M. doi:10.1098/rspa.2000.0713.

215. D. Murugesapillai et al, DNA bridging and looping by HMO1 provides a mechanism for stabilizing nucleosome-free chromatin, Nucl Acids Res (2014) 42 (14): 8996-9004

216. D. Murugesapillai et al, Single-molecule studies of high-mobility group B architectural DNA bending proteins, Biophys Rev (2016) doi:10.1007/s12551-016-0236-4

217. Takenaka, Musashi; Miyachi, Yusuke; Ishii, Jun; Ogino, Chiaki; Kondo, Akihiko (2015-03-04). "The mapping of yeast's G-protein coupled receptor with an atomic force microscope". Nanoscale. 7 (11): 4956–4963. doi:10.1039/c4nr05940a. ISSN 2040-3372.

218. Radmacher, M. (1997). "Measuring the elastic properties of biological samples with the AFM". IEEE Eng Med Biol Mag. 16 (2): 47–57. doi:10.1109/51.582176. PMID 9086372.

219. Perkins, Thomas. "Atomic force microscopy measures properties of proteins and protein folding". SPIE Newsroom. Retrieved 4 March 2016.

220. Thomas, G.; Y. Ouabbas; P. Grosseau; M. Baron; A. Chamayou; L. Galet (2009). "Modeling the mean interaction forces between power particles. Application to silice gel-magnesium stearate mixtures". Applied Surface Science. 255: 7500–7507. Bibcode:2009ApSS..255.7500T. doi:10.1016/j.apsusc.2009.03.099.

221. Sugimoto, Y; Pou, P; Abe, M; Jelinek, P; Pérez, R; Morita, S; Custance, O (Mar 2007). "Chemical identification of individual surface atoms by atomic-force microscopy". Nature. 446 (7131): 64–7. Bibcode:2007Natur.446...64S. doi:10.1038/nature05530. ISSN 0028-0836. PMID 17330040.

222. Bryant, P. J.; Miller, R. G.; Yang, R.; "Scanning tunneling and atomic-force microscopy combined". Applied Physics Letters, Jun 1988, Vol: 52 Issue:26, p. 2233–2235, ISSN 0003-6951.

223. Oscar H. Willemsen, Margot M.E. Snel, Alessandra Cambi, Jan Greve, Bart G. De Grooth and Carl G. Figdor "Biomolecular Interactions Measured by Atomic Force Microscopy" Biophysical Journal, Volume 79, Issue 6, December 2000, Pages 3267-3281.

224. Koo-Hyun Chung and Dae-Eun Kim, "Wear characteristics of diamond-coated atomic force microscope probe". Ultramicroscopy, Volume 108, Issue 1, December 2007, Pages 1-10

225. Xu, Xin; Raman, Arvind (2007). "Comparative dynamics of magnetically, acoustically, and Brownian motion driven microcantilevers in liquids". J. Appl. Phys. 102: 034303. Bibcode:2007JAP...102a4303Y. doi:10.1063/1.2751415.

226. Hasselbach, K.; Ladam, C. (2008). "High resolution magnetic imaging : MicroSQUID Force Microscopy". Journal of Physics: Conference Series. 97: 012330. Bibcode:2008JPhCS..97a2330H. doi:10.1088/1742-6596/97/1/012330.

227. Giessibl, Franz J. (1 January 1998). "High-speed force sensor for force microscopy and profilometry utilizing a quartz tuning fork". Applied Physics Letters. 73 (26): 3956. Bibcode:1998ApPhL..73.3956G. doi:10.1063/1.122948.

228. Nishida, Shuhei; Kobayashi, Dai; Sakurada, Takeo; Nakazawa, Tomonori; Hoshi, Yasuo; Kawakatsu, Hideki (1 January 2008). "Photothermal excitation and laser Doppler velocimetry of higher cantilever vibration modes for dynamic atomic-force microscopy in liquid". Review of Scientific Instruments. 79 (12): 123703. Bibcode:2008RScI...7913703N. doi:10.1063/1.3040500. PMID 19123565.

229. Rugar, D.; Mamin, H. J.; Guethner, P. (1 January 1989). "Improved fiber-optic interferometer for atomic-force microscopy". Applied Physics Letters. 55 (25): 2588. Bibcode:1989ApPhL..55.2588R. doi:10.1063/1.101987.

230. Göddenhenrich, T. "Force microscope with capacitive displacement detection". Journal of Vacuum Science and Technology A. 8 (1): 383. doi:10.1116/1.576401.

231. Giessibl, F. J.; Trafas, B. M. (1 January 1994). "Piezoresistive cantilevers utilized for scanning tunneling and scanning force microscope in ultrahigh vacuum". Review of Scientific Instruments. 65 (6): 1923. Bibcode:1994RScI...65.1923G. doi:10.1063/1.1145232.

232. Binnig, G.; Smith, D. P. E. (1986). "Single-tube three-dimensional scanner for scanning tunneling microscopy". Review of Scientific Instruments. 57 (8): 1688. Bibcode:1986RScI...57.1688B. doi:10.1063/1.1139196. ISSN 0034-6748.

233. R. V. Lapshin (1995). "Analytical model for the approximation of hysteresis loop and its application to the scanning tunneling microscope" (PDF). Review of Scientific Instruments. USA: AIP. 66 (9): 4718–4730. Bibcode:1995RScI...66.4718L. doi:10.1063/1.1145314. ISSN 0034-6748. (Russian translation is available).

234. R. V. Lapshin (2011). "Feature-oriented scanning probe microscopy". In H. S. Nalwa. Encyclopedia of Nanoscience and Nanotechnology (PDF). 14. USA: American Scientific Publishers. pp. 105–115. ISBN 1-58883-163-9.

235. R. V. Lapshin (1998). "Automatic lateral calibration of tunneling microscope scanners" (PDF). Review of Scientific Instruments. USA: AIP. 69 (9): 3268–3276. Bibcode:1998RScI...69.3268L. doi:10.1063/1.1149091. ISSN 0034-6748.

236. R. V. Lapshin (2004). "Feature-oriented scanning methodology for probe microscopy and nanotechnology"(PDF). Nanotechnology. UK: IOP. 15 (9): 1135–1151. Bibcode:2004Nanot..15.1135L. doi:10.1088/0957-4484/15/9/006. ISSN 0957-4484.

237. R. V. Lapshin (2007). "Automatic drift elimination in probe microscope images based on techniques of counter-scanning and topography feature recognition" (PDF). Measurement Science and Technology. UK: IOP. 18 (3): 907–927. Bibcode:2007MeScT..18..907L. doi:10.1088/0957-0233/18/3/046. ISSN 0957-0233.

238. V. Y. Yurov; A. N. Klimov (1994). "Scanning tunneling microscope calibration and reconstruction of real image: Drift and slope elimination" (PDF). Review of Scientific Instruments. USA: AIP. 65 (5): 1551–1557. Bibcode:1994RScI...65.1551Y. doi:10.1063/1.1144890. ISSN 0034-6748.

239. G. Schitter; M. J. Rost (2008). "Scanning probe microscopy at video-rate". Materials Today. UK: Elsevier. 11 (special issue): 40–48. doi:10.1016/S1369-7021(09)70006-9. ISSN 1369-7021. Archived from the original (PDF) on

September 9, 2009.

240. R. V. Lapshin; O. V. Obyedkov (1993). "Fast-acting piezoactuator and digital feedback loop for scanning tunneling microscopes" (PDF). Review of Scientific Instruments. USA: AIP. 64 (10): 2883–2887. Bibcode:1993RScI...64.2883L. doi:10.1063/1.1144377. ISSN 0034-6748.

241. Jong, Wim H De; Borm, Paul JA (June 2008). "Drug Delivery and Nanoparticles: Applications and Hazards". Dovepress International Journal of Nanomedicine: 133–149.

242. Pyrgiotakis, Georgios; Blattmann, Christoph O.; Demokritou, Philip (10 June 2014). "Real-Time Nanoparticle-Cell Interactions in Physiological Media by Atomic Force Microscopy". ACS Sustainable Chemistry & Engineering (Sustainable Nanotechnology 2013): 1681–1690.

243. Evans, Evan A.; Calderwood, David A. (25 May 2007). "Forces and Bond Dynamics in Cell Adhesion". Science. 316 (5828): 1148–1153. Bibcode:2007Sci...316.1148E. doi:10.1126/science.1137592.

244. Scheuring, Simon; Lévy, Daniel; Rigaud, Jean-Louis (1 July 2005). "Watching the Components". Elsevier. 1712 (2): 109–127. doi:10.1016/j.bbamem.2005.04.005.

245. Alsteens, David; Verbelen, Claire; Dague, Etienne; Raze, Dominique; Baulard, Alain R.; Dufrêne, Yves F. (April 2008). "Organization of the Mycobacterial Cell Wall: A Nanoscale View". Pflügers Archive European Journal of Physiology. 456 (1): 117–125. doi:10.1007/s00424-007-0386-0.

CHAPTER 7

Applied Physics in Medicine

Chapter Contents

A medical physicist is a health care professional who applies medical physics to a variety of health-related fields and services. Medical physics is an applied type of physics that involves diagnosing and treating diseases.

INTRODUCTION

Medical physics (also called biomedical physics, medical biophysics or applied physics in medicine) is, generally speaking, the application of physics concepts, theories and methods to medicine or healthcare. Medical physics departments may be found in hospitals or universities.

In the case of hospital work, the term 'Medical Physicist' is the title of a specific healthcare profession with a specific mission statement (see below). Such Medical Physicists are often found in the following healthcare specialties: diagnostic and intervention radiology (also known as medical imaging), nuclear medicine, and radiation oncology (also known as radiotherapy). However, areas of specialty are widely varied in scope and breadth, e.g. clinical physiology (also known as physiological measurement, several countries), neurophysiology (Finland), radiation protection (many countries), and audiology (Netherlands). University departments are of two types. The first type are mainly concerned with preparing students for a career as a hospital medical physicist and research focuses on improving the practice of the profession. A second type (increasingly called 'biomedical physics') has a much wider scope and may include research in any applications of physics to medicine from the study of biomolecular structure to microscopy and nanomedicine. For example, physicist Richard Feynman theorized about the future of nanomedicine. He wrote about the idea of a *medical* use for biological machines (see nanobiotechnology). Feynman and Albert Hibbs suggested that certain repair machines might one day be reduced in size to the point that it would be possible to (as Feynman put it) "swallow the doctor'. The idea was discussed in Feynman's 1959 essay *There's Plenty of Room at the Bottom.*

MISSION STATEMENT OF MEDICAL PHYSICISTS

In the case of hospital medical physics departments, the mission statement is the following:

"Medical Physicists will contribute to maintaining and improving the quality, safety and cost-effectiveness of healthcare services through patient-oriented activities requiring expert action, involvement or advice regarding the specification, selection, acceptance testing, commissioning, quality assurance/control and optimised clinical use of medical devices and regarding patient risks and protection from associated physical agents (e.g., x-rays, electromagnetic fields, laser light, radionuclides) including the prevention of unintended or accidental exposures; all activities will be based on current best evidence or own scientific research when the available evidence is not sufficient. The scope includes risks to volunteers in biomedical research, carers and comforters. The scope often includes risks to workers and public particularly when these impact patient risk"

The term "physical agents" refers to ionising and non-ionising electromagnetic radiations, static electric and magnetic fields, ultrasound, laser light and any other Physical Agent associated with medical e.g., x-rays in computerised tomography (CT), gamma rays/radionuclides in nuclear medicine, magnetic fields and radio-frequencies in magnetic resonance imaging (MRI), ultrasound in ultrasound imaging and Doppler measurements etc.

This mission includes the following 11 key activities:

1. Scientific problem solving service: Comprehensive problem solving service involving recognition of less than optimal performance or optimised use of medical devices, identification and elimination of possible causes or misuse, and confirmation that proposed solutions have restored device performance and use to acceptable status. All activities are to be based on current best scientific evidence or own research when the available evidence is not sufficient.

2. Dosimetry measurements: Measurement of doses suffered by patients, volunteers in biomedical research, carers, comforters and persons subjected to non-medical imaging exposures (e.g., for legal or employment purposes); selection, calibration and maintenance of dosimetry related instrumentation; independent checking of dose related quantities provided by dose reporting devices (including software devices); measurement of dose related quantities required as inputs to dose reporting or estimating devices (including software). Measurements to be based on current recommended techniques and protocols. Includes dosimetry of all physical agents.

3. Patient safety/risk management (including volunteers in biomedical research, carers, comforters and persons subjected to non-medical imaging exposures. Surveillance of medical devices and evaluation of clinical protocols to ensure the ongoing protection of patients, volunteers in biomedical research, carers, comforters and persons subjected to non-medical imaging exposures from the deleterious effects of physical agents in accordance with the latest published evidence or own research when the available evidence is not sufficient. Includes the development of risk assessment protocols.

4. Occupational and public safety/risk management (when there is an impact on medical exposure or own safety). Surveillance of medical devices and evaluation of clinical protocols with respect to protection of workers and public when impacting the exposure of patients, volunteers in biomedical research, carers, comforters and persons subjected to non-medical imaging exposures or responsibility with respect to own safety. Includes the development of risk assessment protocols in conjunction with other experts involved in occupational / public risks.

5. Clinical medical device management: Specification, selection, acceptance testing, commissioning and quality assurance/ control of medical devices in accordance with the latest published European or International recommendations and the management and supervision of associated programmes. Testing to be based on current recommended techniques and protocols.

6. Clinical involvement: Carrying out, participating in and supervising everyday radiation protection and quality control procedures to ensure ongoing effective and optimised use of medical radiological devices and including patient specific optimization.

7. Development of service quality and cost-effectiveness: Leading the introduction of new medical radiological devices into clinical service, the introduction of new medical physics services and participating in the introduction/development of clinical protocols/techniques whilst giving due attention to economic issues.

8. Expert consultancy: Provision of expert advice to outside clients (e.g., clinics with no in-house medical physics expertise).

9. Education of healthcare professionals (including medical physics trainees: Contributing to quality healthcare professional education through knowledge transfer activities concerning the technical-scientific knowledge, skills and competences supporting the clinically effective, safe, evidence-based and economical use of medical radiological devices. Participation in the education of medical physics students and organisation of medical physics residency programmes.

10. Health technology assessment (HTA): Taking responsibility for the physics component of health technology assessments related to medical radiological devices and /or the medical uses of radioactive substances/sources.

11. Innovation: Developing new or modifying existing devices (including software) and protocols for the solution of hitherto unresolved clinical problems.

Medical Biophysics and Biomedical Physics

Some education institutions house departments or programs bearing the title "medical biophysics" or "biomedical physics" or "applied physics in medicine". Generally, these fall into one of two categories: interdisciplinary departments that house biophysics, radiobiology, and medical physics under a single umbrella; and undergraduate programs that prepare students for further study in medical physics, biophysics, or medicine.

AREAS OF SPECIALITY

The International Organization for Medical Physics (IOMP) - the world's premier professional organization for medical physics with nearly 22,000 members in 84 countries - recognizes main areas of medical physics employment and focus. These are:

Medical Imaging Physics

Para-sagittal MRI of the head in a patient with benign familial macrocephaly.

Medical imaging physics is also known as diagnostic and interventional radiology physics. Clinical (both "in-house" and "consulting") physicists typically deal with areas of testing, optimization, and quality assurance of diagnostic radiology physics areas such as radiographic X-rays, fluoroscopy, mammography, angiography, and computed tomography, as well as non-ionizing radiation modalities such as ultrasound, and MRI. They may also be engaged with radiation protection issues such as radiation exposure monitoring and dosimetry. In addition, many imaging physicists are often also involved with nuclear medicine systems, including single photon emission computed tomography (SPECT) and positron emission tomography (PET). Sometimes, imaging physicists may be engaged in clinical areas, but for research and teaching purposes, such as e.g. quantifying intravascular ultrasound as a possible method of imaging a particular vascular object.

Radiation Therapeutic Physics

Radiation therapeutic physics is also known as radiotherapy physics or radiation oncology physics. The majority of medical physicists currently working in the US, Canada, and some western countries are of this group. A Radiation Therapy physicist typically deals with linear accelerator (Linac) systems and kilovoltage x-ray treatment units on a daily basis, as well as more advanced modalities such as TomoTherapy, gamma knife, cyberknife, proton therapy, and brachytherapy. The academic and research side of therapeutic physics may encompass fields such as boron neutron capture therapy, sealed source radiotherapy, terahertz radiation, high-intensity focused ultrasound (including lithotripsy), optical radiation lasers, ultraviolet etc. including photodynamic therapy, as well as nuclear medicine including unsealed source radiotherapy, and photomedicine, which is the use of light to treat and diagnose disease.

Nuclear Medicine Physics

This is a branch of medicine that uses radiation to provide information about the functioning of a person's specific organs or to treat disease. In most cases, the information is used by physicians to make a quick, accurate diagnosis of the patient's illness. The thyroid, bones, heart, liver and many other organs can be easily imaged, and disorders in their function revealed. In some cases radiation can be used to treat diseased organs, or tumours. Five Nobel Laureates have been intimately involved with the use of radioactive tracers in medicine. Over 10,000 hospitals worldwide use radioisotopes in medicine, and about 90% of the procedures are for diagnosis. The most common radioisotope used in diagnosis is technetium-99, with some 30 million procedures per year, accounting for 80% of all nuclear medicine procedures worldwide. In developed countries (26% of world population) the frequency of diagnostic nuclear medicine is 1.9% per year, and the frequency of therapy with radioisotopes is about one tenth of this. In the USA there are some 18 million nuclear medicine procedures per year among 311 million people, and in Europe about 10 million among 500 million people. In Australia there are about 560,000 per year among 21 million people, 470,000 of these using reactor isotopes. The use of radiopharmaceuticals in diagnosis is growing at over 10% per year. Nuclear medicine was developed in the 1950s by physicians with an endocrine emphasis, initially using iodine-131 to diagnose and then treat thyroid disease. In recent years, specialists have also come from radiology, as dual CT/PET procedures have become established. Computed X-ray tomography (CT) scans and nuclear medicine contribute 36% of the total radiation exposure and 75% of the medical exposure to the US population, according to a US National Council on Radiation Protection & Measurements report in 2009. The report showed that Americans' average total yearly radiation exposure had increased from 3.6 millisievert to 6.2 mSv per year since the early 1980s, due to medical-related procedures. (Industrial radiation exposure, including that from nuclear power plants, is less than 0.1% of overall public radiation exposure.)

Health Physics

Health physics is also known as Physical Hazards or Physical Safety or Radiation Safety or Radiation Protection.

- Background radiation
- Radiation protection
- Dosimetry
- Health physics
- Radiological protection of patients

Clinical Audiology Physics

- Physics and acoustics for audiology: the basic physics of sound, instrumentation, and the principles of digital signal processing involved in audiological research. Topics include the physics of sound waves, room acoustics, the measurement of reverberation time; the nature of acoustic impedance; the nature of filters and amplifiers, acoustics of speech, calibration.

Laser Medicine

- Lasers and applications in medicine

Medical Optics

- Optical imaging

Neurophysics

Neurophysics (or neural physics) is also known as functional neuroimaging. It refers to imaging structure and function in the nervous system as well as:

- fMRI
- molecular imaging
- electrical impedance tomography
- diffuse optical imaging
- optical coherence tomography
- dual energy X-ray absorptiometry
- image-guided surgery

Cardiophysics

- Cardiophysics (or cardiovascular physics) is using the methods of, and theories from, physics to study cardiovascular system at different levels of its organisation, from the molecular scale to whole organisms.
- Electrocardiography

Physiological Measurement Techniques

Physiological measurements have also been used to monitor and measure various physiological parameters. Many physiological measurement techniques are non-invasive and can be used in conjunction with, or as an alternative to, other invasive methods.

- Electroencephalography
- Electromyography
- Electronystagmography
- Endoscopy
- Medical ultrasonography
- Non-ionising radiation (lasers, ultraviolet etc.)
- Near infrared spectroscopy

- Pulse oximetry
- Blood gas monitor
- Blood pressure measurement

Physics of the Human and Animal Bodies

8.1 Biomechanics

- Body mass index
- Sports biomechanics
- Biofluid mechanics

8.2 Bioelectromagnetics of the human and animal bodies

- Bioelectrodynamics
- Biomagnetism
- Electromagnetic radiation and health
- Electrophysiology

8.3 Biomaterials and artificial organs

- Biomaterials for cell and organ therapies
- Cardiovascular biomaterials, artificial heart and cardiac assist devices
- Artificial skin, bones, joints, teeth and related biomaterials
- Histopathology
- Drug delivery and control release

Healthcare informatics and Computational Physics

- Information and communication in medicine
- Medical informatics
- Image processing, display and visualization
- Computer-aided diagnosis
- Picture archiving and communication systems (PACS)
- Standards: DICOM, ISO, IHE
- Hospital information systems
- e-Health
- Telemedicine
- Digital operating room
- Workflow, patient-specific modeling
- Medicine on the Internet of Things
- Distant monitoring and telehomecare

Areas of Research and Academic Development

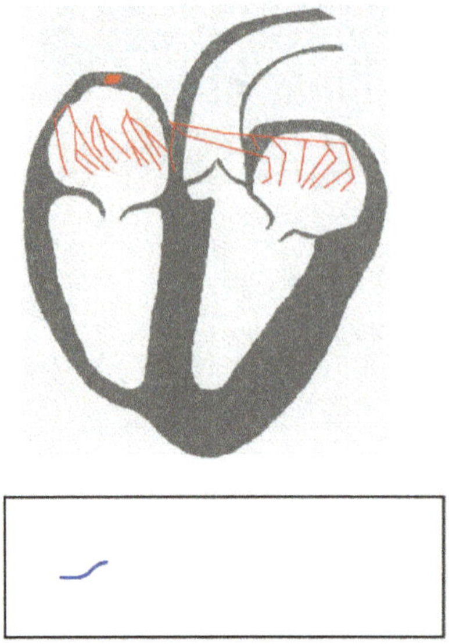

ECG trace

Non-clinical physicists may or may not focus on the above areas from an academic and research point of view, but their scope of specialization may also encompass lasers and ultraviolet systems (such as photodynamic therapy), fMRI and other methods for functional imaging as well as molecular imaging, electrical impedance tomography, diffuse optical imaging, optical coherence tomography, and dual energy X-ray absorptiometry.

EDUCATION AND TRAINING

In Australia and New Zealand

The Australasian College of Physical Scientists and Engineers in Medicine (ACPSEM) is the professional body that oversees the education and certification of medical physicists in Australia and New Zealand and has a mission to advance services and professional standards in medical physics and biomedical engineering.

In Europe

The presence of Medical Physicists at Expert level ('Medical Physics Experts') in healthcare in Europe is required by EC Directive 2013/59/EURATOM. The European Federation of Organizations for Medical Physics (EFOMP) has defined a detailed inventory of learning outcomes for Medical Physics Experts in terms of Knowledge, Skills and Competences (the latter in Europe means 'responsibilities'). In Europe the professional preparation for

Medical Physicists consists of a first degree in Physics or equivalent (e.g., biophysics, electrical or mechanical engineering), a Masters in Medical Physics and a 2-year training Residency. For the latest EFOMP policy statement on the role and education and training requirements of the Medical Physicist and Medical Physics Expert go to. For the final version of the EC funded document 'European Guidelines on the Medical Physics Expert' download Report 174 and its annexes from.

In North America

In North America, medical physics training is offered at the master's, doctorate, post-doctorate and/or residency levels. Also, a professional doctorate has been recently introduced as an option. Several universities offer these degrees in Canada and the United States.

As of October 2013, over 70 universities in North America have medical physics graduate programs or residencies that are accredited by *The Commission on Accreditation of Medical Physics Education Programs* (CAMPEP). The majority of residencies are therapy, but diagnostic and nuclear are also on the rise in the past several years.

Professional certification is obtained from the American Board of Radiology (for all 4 areas), the American Board of Medical Physics (for MRI), the American Board of Science in Nuclear Medicine (for Nuc Med and PET), and the Canadian College of Physicists in Medicine. As of 2012, enrollment in a CAMPEP-accredited residency or graduate program is required to start the ABR certification process. Starting in 2014, completion of a CAMPEP-accredited residency will be required to advance to part 2 of the ABR certification process.

United Kingdom

From October 2011 as part of the Modernising Scientific Careers scheme, the route to accreditation as a medical physicist in England and Wales is provided by the Scientist Training Programme (STP). This scheme is a three-year graduate scheme provided by the National School of Healthcare Science. Entrants are required to have a prior undergraduate degree (1:1 or 2:1) in an appropriate physical science.

The STP involves a part-time MSc in Medical Physics (provided by either King's College London, University of Liverpool or University of Newcastle) in addition to practical training within the National Health Service. Assessment is provided by the completion of competencies and by a final assessment similar to the OSCE undertaken by other clinical staff. Completion of the STP leads to accreditation with the Institute of Physics and Engineering in Medicine (IPEM) and registration as a Clinical Scientist.

Prior to 2011 the training route in the United Kingdom was administered in two parts, and this scheme is still used in Scotland and Northern Ireland). Part I involved limited clinical experience and a full-time MSc in medical physics. Part II involved exclusively clinical experience in which the candidate would produce a portfolio of experience which would be submitted to the Academy for Healthcare Science which (in addition to a viva) would lead to professional accreditation with IPEM.

International

There are regular regional and international educational medical physics activities. The oldest of these is the International College on Medical Physics at the International Centre for Theoretical Physics (ICTP), Trieste, Italy. This College has educated more than 1000 medical physicists from developing countries.

Legislative and Advisory Bodies

- ICRU: International Commission on Radiation Units and Measurements
- ICRP: International Commission on Radiological Protection
- NCRP: National Council on Radiation Protection & Measurements
- NRC: Nuclear Regulatory Commission
- FDA: Food and Drug Administration
- IAEA: International Atomic Energy Agency

MOLECULAR IMAGING

Molecular imaging originated from the field of radiopharmacology due to the need to better understand fundamental molecular pathways inside organisms in a noninvasive manner.

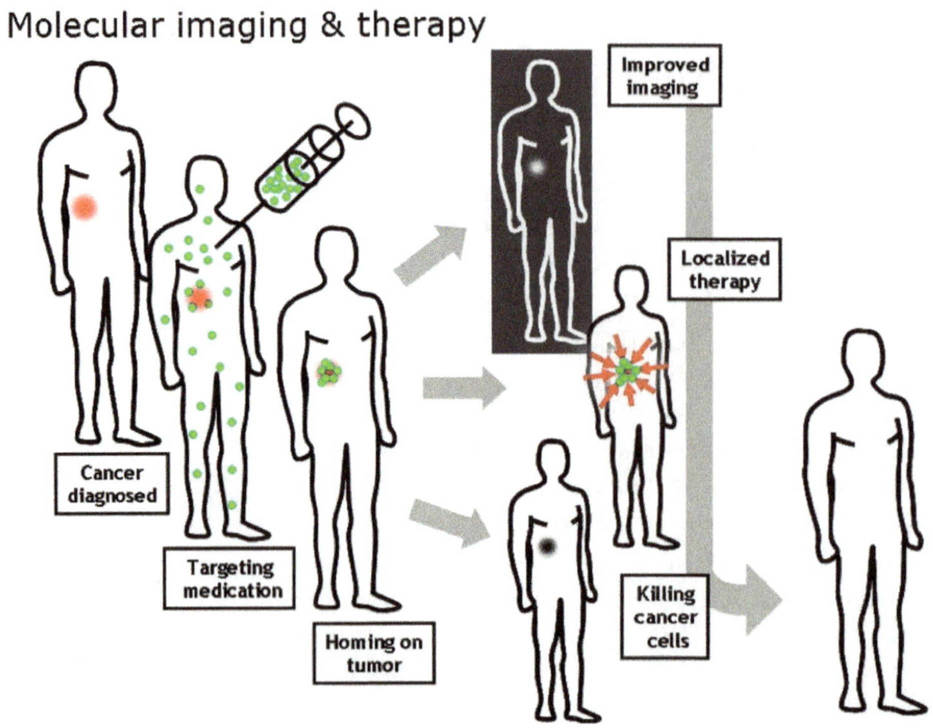

Molecular Imaging emerged in the early twenty-first century as a discipline at the intersection of molecular biology and *in vivo* imaging. It enables the visualisation of the cellular function and the follow-up of the molecular process in living organisms without perturbing them. The multiple and numerous potentialities of this field are applicable to the diagnosis of diseases such as cancer, and neurological and cardiovascular diseases. This technique also contributes to improving the treatment of these disorders by optimizing the pre-clinical and clinical tests of new medication. They are also expected to have a major economic impact due to earlier and more precise diagnosis. Molecular and Functional Imaging has taken on a new direction since the description of the human genome. New paths in fundamental research, as well as in applied and industrial research, render the task of scientists more complex and increase the demands on them. Therefore, a tailor-made teaching program is in order.

Molecular imaging differs from traditional imaging in that probes known as biomarkers are used to help image particular targets or pathways. Biomarkers interact chemically with their surroundings and in turn alter the image according to molecular changes occurring within the area of interest. This process is markedly different from previous methods of imaging which primarily imaged differences in qualities such as density or water content. This ability to image fine molecular changes opens up an incredible number of exciting possibilities for medical application, including early detection and treatment of disease and basic pharmaceutical development. Furthermore, molecular imaging allows for quantitative tests, imparting a greater degree of objectivity to the study of these areas. One emerging technology is MALDI molecular imaging based on mass spectrometry.

Many areas of research are being conducted in the field of molecular imaging. Much research is currently centered on detecting what is known as a predisease state or molecular states that occur before typical symptoms of a disease are detected. Other important veins of research are the imaging of gene expression and the development of novel biomarkers. Organizations such as the SNMMI Center for Molecular Imaging Innovation and Translation (CMIIT) have formed to support research in this field. In Europe, other "networks of excellence" such as DiMI (Diagnostics in Molecular Imaging) or EMIL (European Molecular Imaging Laboratories) work on this new science, integrating activities and research in the field. In this way, a European Master Programme "EMMI' is being set up to train a new generation of professionals in molecular imaging.

Recently the term "Molecular Imaging" has been applied to a variety of microscopy and nanoscopy techniques including live-cell microscopy, Total Internal Reflection Fluorescence (TIRF)-microscopy, STimulated Emission Depletion (STED)-nanoscopy and Atomic Force Microscopy (AFM) as here images of molecules are the readout.

Imaging Modalities

There are many different modalities that can be used for noninvasive molecular imaging. Each have their different strengths and weaknesses and some are more adept at imaging multiple targets than others.

Magnetic Resonance Imaging

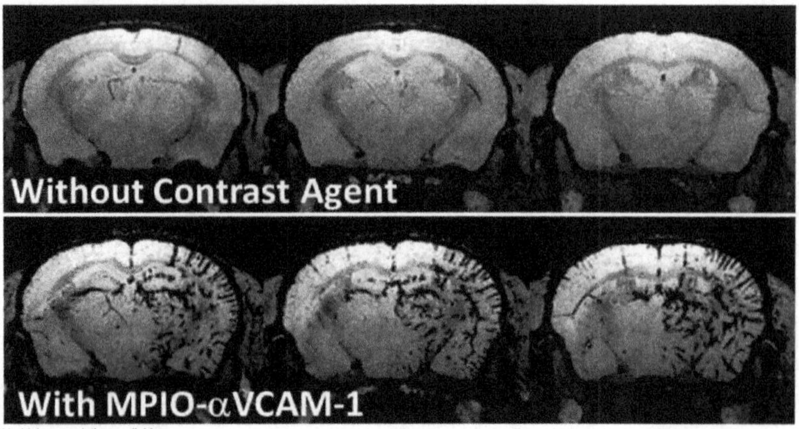

Molecular MRI of a mouse brain presenting acute inflammation in the right hemisphere. Whereas unenhanced MRI failed to reveal any difference between right en left hemispheres, injection of a contrast-agent targeted to inflamed vessels allows to reveal inflammation specifically in the right hemisphere.

MRI has the advantages of having very high spatial resolution and is very adept at morphological imaging and functional imaging. MRI does have several disadvantages though. First, MRI has a sensitivity of around 10^{-3} mol/L to 10^{-5} mol/L which, compared to other types of imaging, can be very limiting. This problem stems from the fact that the difference between atoms in the high energy state and the low energy state is very small. For example, at 1.5 teslas, a typical field strength for clinical MRI, the difference between high and low energy states is approximately 9 molecules per 2 million. Improvements to increase MR sensitivity include increasing magnetic field strength, and hyperpolarization via optical pumping, dynamic nuclear polarization or parahydrogen induced polarization. There are also a variety of signal amplification schemes based on chemical exchange that increase sensitivity.

To achieve molecular imaging of disease biomarkers using MRI, targeted MRI contrast agents with high specificity and high relaxivity (sensitivity) are required. To date, many studies have been devoted to developing targeted-MRI contrast agents to achieve molecular imaging by MRI. Commonly, peptides, antibodies, or small ligands, and small protein domains, such as HER-2 affibodies, have been applied to achieve targeting. To enhance the sensitivity of the contrast agents, these targeting moieties are usually linked to high payload MRI contrast agents or MRI contrast agents with high relaxivities. In particular, the recent development of micron-sized particles of iron oxide (MPIO) allowed to reach unprecedented levels of sensitivity to detect proteins expressed by arteries and veins.

Optical Imaging

There are a number of approaches used for optical imaging. The various methods depend upon fluorescence, bioluminescence, absorption or reflectance as the source of contrast.

Optical imaging's most valuable attribute is that it and ultrasound do not have strong safety concerns like the other medical imaging modalities.

The downside of optical imaging is the lack of penetration depth, especially when working at visible wavelengths. Depth of penetration is related to the absorption and scattering of light, which is primarily a function of the wavelength of the excitation source. Light is absorbed by endogenous chromophores found in living tissue (e.g. hemoglobin, melanin, and lipids). In general, light absorption and scattering decreases with increasing wavelength. Below ~700 nm (e.g. visible wavelengths), these effects result in shallow penetration depths of only a few millimeters. Thus, in the visible region of the spectrum, only superficial assessment of tissue features is possible. Above 900 nm, water absorption can interfere with signal-to-background ratio. Because the absorption coefficient of tissue is considerably lower in the near infrared (NIR) region (700-900 nm), light can penetrate more deeply, to depths of several centimeters.

Near Infrared Imaging

Fluorescent probes and labels are an important tool for optical imaging. Some researchers have applied NIR imaging in Rat model of acute myocardial infarction (AMI), using a peptide probe that can binds to apoptotic and necrotic cells. A number of near-infrared (NIR) fluorophores have been employed for in vivo imaging, including Kodak X-SIGHT Dyes and Conjugates, Pz 247, DyLight 750 and 800 Fluors, Cy 5.5 and 7 Fluors, Alexa Fluor 680 and 750 Dyes, IRDye 680 and 800CW Fluors. Quantum dots, with their photostability and bright emissions, have generated a great deal of interest; however, their size precludes efficient clearance from the circulatory and renal systems while exhibiting long-term toxicity..

Several studies have demonstrated the use of infrared dye-labeled probes in optical imaging.

- In a comparison of gamma scintigraphy and NIR imaging, a cyclopentapeptide dual-labeled with 111 In and an NIR fluorophore was used to image αvβ3-integrin positive melanoma xenografts.
- Near-infrared labeled RGD targeting αvβ3-integrin has been used in numerous studies to target a variety of cancers.
- An NIR fluorophore has been conjugated to epidermal growth factor (EGF) for imaging of tumor progression.
- An NIR fluorophore was compared to Cy5.5, suggesting that longer-wavelength dyes may produce more effective targeting agents for optical imaging.
- Pamidronate has been labeled with an NIR fluorophore and used as a bone imaging agent to detect osteoblastic activity in a living animal.
- An NIR fluorophore-labeled GPI, a potent inhibitor of PSMA (prostate specific membrane antigen).
- Use of human serum albumin labeled with an NIR fluorophore as a tracking agent for mapping of sentinel lymph nodes.
- 2-Deoxy-D-glucose labeled with an NIR fluorophore.

It is important to note that addition of an NIR probe to any vector can alter the vector's biocompatibility and biodistribution. Therefore, it can not be unequivocally assumed that the conjugated vector will behave similarly to the native form.

Single Photon Emission Computed Tomography

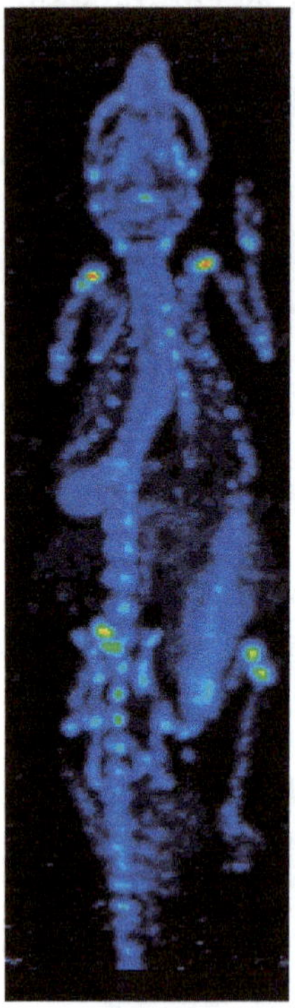

SPECT image (bone tracer) of a mouse MIP

The development of computed tomography in the 1970s allowed mapping of the distribution of the radioisotopes in the organ or tissue, and led to the technique now called single photon emission computed tomography (SPECT).

The imaging agent used in SPECT emits gamma rays, as opposed to the positron emitters (such as ^{18}F) used in PET. There are a range of radiotracers (such as ^{99m}Tc, ^{111}In, ^{123}I, ^{201}Tl) that can be used, depending on the specific application.

Xenon (^{133}Xe) gas is one such radiotracer. It has been shown to be valuable for diagnostic inhalation studies for the evaluation of pulmonary function; for imaging the lungs; and may also be used to assess rCBF. Detection of this gas occurs via a gamma camera—which is a scintillation detector consisting of a collimator, a NaI crystal, and a set of photomultiplier tubes.

By rotating the gamma camera around the patient, a three-dimensional image of the distribution of the radiotracer can be obtained by employing filtered back projection or other tomographic techniques. The radioisotopes used in SPECT have relatively long half lives (a few hours to a few days) making them easy to produce and relatively cheap.

This represents the major advantage of SPECT as a molecular imaging technique, since it is significantly cheaper than either PET or fMRI. However it lacks good spatial (i.e., where exactly the particle is) or temporal (i.e., did the contrast agent signal happen at this millisecond, or that millisecond) resolution. Additionally, due to the radioactivity of the contrast agent, there are safety aspects concerning the administration of radioisotopes to the subject, especially for serial studies.

Positron Emission Tomography

Positron emission tomography (PET) is a nuclear medicine imaging technique which produces a three-dimensional image or picture of functional processes in the body. The theory behind PET is simple enough. First a molecule is tagged with a positron emitting isotope. These positrons annihilate with nearby electrons, emitting two 511 keV photons, directed 180 degrees apart in opposite directions. These photons are then detected by the scanner which can estimate the density of positron annihilations in a specific area. When enough interactions and annihilations have occurred, the density of the original molecule may be measured in that area. Typical isotopes include ^{11}C, ^{13}N, ^{15}O, ^{18}F, ^{64}Cu, ^{62}Cu, ^{124}I, ^{76}Br, ^{82}Rb, ^{89}Zr and ^{68}Ga, with ^{18}F being the most clinically utilized. One of the major disadvantages of PET is that most of the probes must be made with a cyclotron. Most of these probes also have a half life measured in hours, forcing the cyclotron to be on site. These factors can make PET prohibitively expensive. PET imaging does have many advantages though. First and foremost is its sensitivity: a typical PET scanner can detect between 10^{-11} mol/L to 10^{-12} mol/L concentrations.

Probes and the Imaging of Molecular Interactions

In order to image multiple targets you must first identify and develop the interactions of which you are trying to take advantage. Developing good probes is often difficult and is an area of intense research.

REFERENCES

1. Richard P. Feynman (December 1959). "There's Plenty of Room at the Bottom". Retrieved March 2010. Check date values in: |access-date= (help)
2. Guibelalde E., Christofides S., Caruana C. J., Evans S. van der Putten W. (2012). Guidelines on the Medical Physics Expert' a project funded by the European Commission
3. Caruana C.J., Christofides S., Hartmann G.H. (2014) European Federation of Organisations for Medical Physics (EFOMP) Policy Statement 12.1: Recommendations on Medical Physics Education and Training in Europe 2014 Physica Medica - European Journal of Medical Physics, 30:6, p598-603
4. "Department of Medical Biophysics". utoronto.ca.
5. "Medical Biophysics - Western University". uwo.ca.
6. UCLA Biomedical Physics Graduate Program

7. "Welcome". wayne.edu.
8. "Medical Physics". fresnostate.edu.
9. "AAPM Position Statements, Policies and Procedures - Details". aapm.org.
10. "AAPM - What do Medical Physicists Do?". aapm.org.
11. http://www.medphys.ca/content.php?sec=1
12. Hill R, Healy B, Holloway L, Kuncic Z, Thwaites D, Baldock C (2014). "Advances in kilovoltage x-ray beam dosimetry". Physics in Medicine and Biology. 59 (6). Bibcode:2014PMB....59R.183H. doi:10.1088/0031-9155/59/6/R183. PMID 24584183.
13. Thwaites DI, Tuohy JB (2006). "Back to the future: the history and development of the clinical linear accelerator". Physics in Medicine and Biology. 51 (13). Bibcode:2006PMB....51R.343T. doi:10.1088/0031-9155/51/13/R20. PMID 16790912.
14. Mackie, T R (2006). "The history of tomotherapy". Physics in Medicine and Biology. 51 (13). Bibcode:2006PMB....51R.427M. doi:10.1088/0031-9155/51/13/R24.
15. http://www.physicamedica.com/article/S1120-1797(14)00103-3/fulltext
16. https://ec.europa.eu/energy/en/publications-medical-exposure
17. How does someone become a Medical Physicist?. AAPM. Retrieved on 2011-06-25.
18. "CAMPEP Accredited Professional Doctorate Programs in Medical Physics". CAMPEP. CAMPEP. Retrieved 2015-03-29.
19. CAMPEP Accredited Graduate Programs in Medical Physics. Campep.org (2011-06-01). Retrieved on 2011-06-25.
20. IC RP CAMPEP addendum. Theabr.org. Retrieved on 2011-06-25.
21. . Prospects Graduate Website. Retrieved 2014-03-18
22. . NHS Careers Website. Retrieved 2014-03-18.

CHAPTER 8

Bioenergetics

Chapter Contents

Bioenergetics is a multidisciplinary field involving biology, chemistry, and biochemistry that is concerned with the flow of energy through organisms at the cellular level. Different organisms undergo many different processes in the production of energy ranging from photosynthesis in plants to fiber feeding in some protists. All of these lead to cellular respiration, also known as oxidative metabolism, through which individual cells are able to produce usable energy. There are thousands of processes that eventually lead to the production of usable energy in the cells of organisms. Bioenergetics is concerned with understanding how these processes work and how they interact with all of the other energy-production processes in the organism.

INTRODUCTION

Bioenergetics is a field in biochemistry and cell biology that concerns energy flow through living systems. This is an active area of biological research that includes the study of the transformation of energy in living organisms and the study of thousands of different cellular processes such as cellular respiration and the many other metabolic processes that can lead to production and utilization of energy in forms such as ATP molecules.

OVERVIEW

Bioenergetics is the part of biochemistry concerned with the energy involved in making and breaking of chemical bonds in the molecules found in biological organisms. It can also be defined as the study of energy relationships and energy transformations in living organisms.

Growth, development and metabolism are some of the central phenomena in the study of biological organisms. The role of energy is fundamental to such biological processes. The ability to harness energy from a variety of metabolic pathways is a property of all living organisms. Life is dependent on energy transformations; living organisms survive because of exchange of energy within and without.

In a living organism, chemical bonds are broken and made as part of the exchange and transformation of energy. Energy is available for work (such as mechanical work) or for other processes (such as chemical synthesis and anabolic processes in growth), when weak bonds are broken and stronger bonds are made. The production of stronger bonds allows release of usable energy.

Living organisms obtain energy from organic and inorganic materials. For example, lithotrophs can oxidize minerals such as nitrates or forms of sulfur, such as elemental sulfur, sulfites, and hydrogen sulfide to produce ATP. In photosynthesis, autotrophs can produce ATP using light energy. Heterotrophs must consume organic compounds. These are mostly carbohydrates, fats, and proteins. The amount of energy actually obtained by the organism is lower than the amount present in the food; there are losses in digestion, metabolism, and thermogenesis.

The materials are generally combined with oxygen to release energy, although some can also be oxidized anaerobically by various organisms. The bonds holding the molecules of nutrients together and in particular the bonds holding molecules of free oxygen together are relatively weak compared with the chemical bonds holding carbon dioxide and water together. The utilization of these materials is a form of slow combustion. That is why the energy content of food can be estimated with a bomb calorimeter. The materials are oxidized slowly enough that the organisms do not actually produce fire. The oxidation releases energy because stronger bonds have been formed. This net energy may evolve as heat, or some of which may be used by the organism for other purposes, such as breaking other bonds to do chemistry.

Living organisms produce ATP from energy sources via oxidative phosphorylation. The terminal phosphate bonds of ATP are relatively weak compared with the stronger bonds formed when ATP is broken down to adenosine monophosphate and phosphate and then dissolved in water. Here it is the energy of hydration that results in energy release. An organism›s stockpile of ATP is used as a battery to store energy in cells, for intermediate metabolism. Utilization of chemical energy from such molecular bond rearrangement powers biological processes in every biological organism.

Types of Reactions

- Exergonic is a spontaneous reaction that releases energy. It is thermodynamically favored. On the course of a reaction, energy needs to be put in, this activation energy drives the reactants from a stable state to a highly energetic unstable configuration. These reactants are usually complex molecules that are broken into simpler products. The entire reaction is usually catabolic. The release of energy (called Gibbs free energy) is negative and equal to - ΔG because energy is lost from the bonds formed by the products.

- **Endergonic** is an anabolic reaction that consumes energy. It has a positive ΔG because energy is required to break bonds.

The free energy (ΔG) gained or lost in a reaction can be calculated: $\Delta G = \Delta H - T\Delta S$ where G = Gibbs free energy, H = enthalpy, T = temperature, and S = entropy.

Cotransport

In August 1960, Robert K. Crane presented for the first time his discovery of the sodium-glucose cotransport as the mechanism for intestinal glucose absorption. Crane's discovery of cotransport was the first ever proposal of flux coupling in biology and was the most important event concerning carbohydrate absorption in the 20th century.

Chemiosmotic Theory

One of the major triumphs of bioenergetics is Peter D. Mitchell's chemiosmotic theory of how protons in aqueous solution function in the production of ATP in cell organelles such as mitochondria. This work earned Mitchell the 1978 Nobel Prize for Chemistry. Other cellular sources of ATP such as glycolysis were understood first, but such processes for direct coupling of enzyme activity to ATP production are not the major source of useful chemical energy in most cells. Chemiosmotic coupling is the major energy producing process in most cells, being utilized in chloroplasts and several single celled organisms in addition to mitochondria.

Energy Balance

Energy homeostasis is the homeostatic control of energy balance – the difference between energy obtained through food consumption and energy expenditure – in living systems.

ENERGY HOMEOSTASIS

In biology, energy homeostasis, or the homeostatic control of energy balance, is a biological process that involves the coordinated homeostatic regulation of food intake (energy inflow) and energy expenditure (energy outflow). The human brain, particularly the hypothalamus, plays a central role in regulating energy homeostasis and generating the sense of hunger by integrating a number of biochemical signals that transmit information about energy balance. Fifty percent of the energy from glucose metabolism is immediately converted to heat.

Energy homeostasis is an important aspect of bioenergetics.

In the US, biological energy is expressed using the energy unit Calorie with a capital C (i.e. kilocalorie), which equals the energy needed to increase the temperature of 1 kilogram of water by 1 °C (about 4.18 kJ).

Energy balance, through biosynthetic reactions, can be measured with the following equation:

Energy intake (food) = Energy expended (heat + work) + Energy stored.

The first law of thermodynamics states that energy can be neither created nor destroyed. But energy can be converted from one form of energy to another. So when a calorie of food eaten enters a body, ultimately 100% of that calorie will be converted to heat, resulting in three particular short-term effects: a portion of that calorie is either stored as fat, transferred to the body's cells as chemical energy, by Adenosine triphosphate (ATP), a coenzyme and used for mechanical work or chemical synthesis, or immediately dissipated through heat.

ENERGY

Intake

Energy intake is calories consumed as food, which is determined by hunger (hypothalamus) and choice (cerebral cortex). Hunger is determined by the impact of the hormones leptin and ghrelin on the hypothalamus.

Expenditure

Energy expenditure is mainly a sum of internal heat produced and external work. The internal heat produced is, in turn, mainly a sum of basal metabolic rate (BMR) and the thermic effect of food. External work may be estimated by measuring the physical activity level (PAL).

Imbalance

Positive Balance

A positive balance is a result of *energy intake* being *higher* than what is consumed in external work and other bodily means of energy expenditure.

The main preventable causes are:

- Overeating, resulting in increased energy intake
- Sedentary lifestyle, resulting in decreased energy expenditure through external work

A positive balance results in energy being stored as fat and/or muscle, causing weight gain. In time, overweight and obesity may develop, with resultant complications.

Negative Balance

A negative balance is a result of energy intake being less than what is consumed in external work and other bodily means of energy expenditure.

The main cause is *undereating* due to a medical condition such as decreased appetite, anorexia nervosa, digestive disease, or due to some circumstance such as fasting or lack of access to food. Hyperthyroidism can also be a cause.

Requirement

Normal energy requirement, and therefore normal energy intake, depends mainly on age, sex and physical activity level (PAL). The Food and Agriculture Organization (FAO) of the United Nations has compiled a detailed report on human energy requirements: Human energy requirements (Rome, 17–24 October 2001) An older but commonly used and fairly accurate method is the Harris-Benedict equation.

Yet, there are currently ongoing studies to show if calorie restriction to below normal values have beneficial effects, and even though they are showing positive indications in primates it is still not certain if calorie restriction has a positive effect on longevity for primates and humans. Calorie restriction may be viewed as attaining energy balance at a lower intake and expenditure, and is, in this sense, not generally an energy imbalance, except for an initial imbalance where decreased expenditure hasn't yet matched the decreased intake.

REFERENCES

1. FAO, Calculation of the Energy Content of Foods—Energy Conversion Factors
2. Schmidt-Rohr K (2015). "Why Combustions Are Always Exothermic, Yielding About 418 kJ per Mole of O2". J. Chem. Educ. 92 (12): 2094–2099. doi:10.1021/acs.jchemed.5b00333.
3. Robert K. Crane, D. Miller and I. Bihler. "The restrictions on possible mechanisms of intestinal transport of sugars". In: Membrane Transport and Metabolism. Proceedings of a Symposium held in Prague, August 22–27, 1960. Edited by A. Kleinzeller and A. Kotyk. Czech Academy of Sciences, Prague, 1961, pp. 439-449.
4. Wright, Ernest M.; Turk, Eric (2004). "The sodium glucose cotransport family SLC5" (PDF). Pflügers Arch. 447 (5): 510–8. doi:10.1007/s00424-003-1063-6. PMID 12748858. Crane in 1961 was the first to formulate the cotransport concept to explain active transport . Specifically, he proposed that the accumulation of glucose in the intestinal epithelium across the brush border membrane was coupled to downhill Na+
5. transport cross the brush border. This hypothesis was rapidly tested, refined and extended [to] encompass the active transport of a diverse range of molecules and ions into virtually every cell type.

6. Boyd, C A R (2008). "Facts, fantasies and fun in epithelial physiology". Experimental Physiology. 93 (3): 304. doi:10.1113/expphysiol.2007.037523. PMID 18192340. the insight from this time that remains in all current text books is the notion of Robert Crane published originally as an appendix to a symposium paper published in 1960 (Crane et al. 1960). The key point here was 'flux coupling', the cotransport of sodium and glucose in the apical membrane of the small intestinal epithelial cell. Half a century later this idea has turned into one of the most studied of all transporter proteins (SGLT1), the sodium–glucose cotransporter.

7. Peter Mitchell (1961). "Coupling of phosphorylation to electron and hydrogen transfer by a chemi-osmotic type of mechanism". Nature. 191 (4784): 144–8. Bibcode:1961Natur.191..144M. doi:10.1038/191144a0. PMID 13771349.

8. Malenka RC, Nestler EJ, Hyman SE (2009). Sydor A, Brown RY, ed. Molecular Neuropharmacology: A Foundation for Clinical Neuroscience (2nd ed.). New York: McGraw-Hill Medical. pp. 179, 262–263. ISBN 9780071481274. Orexin neurons are regulated by peripheral mediators that carry information about energy balance, including glucose, leptin, and ghrelin. ... Accordingly, orexin plays a role in the regulation of energy homeostasis, reward, and perhaps more generally in emotion. ... The regulation of energy balance involves the exquisite coordination of food intake and energy expenditure. Experiments in the 1940s and 1950s showed that lesions of the lateral hypothalamus (LH) reduced food intake; hence, the normal role of this brain area is to stimulate feeding and decrease energy utilization. In contrast, lesions of the medial hypothalamus, especially the ventromedial nucleus (VMH) but also the PVN and dorsomedial hypothalamic nucleus (DMH), increased food intake; hence, the normal role of these regions is to suppress feeding and increase energy utilization. Yet discovery of the complex networks of neuropeptides and other neurotransmitters acting within the hypothalamus and other brain regions to regulate food intake and energy expenditure began in earnest in 1994 with the cloning of the leptin (ob, for obesity) gene. Indeed, there is now explosive interest in basic feeding mechanisms given the epidemic proportions of obesity in our society, and the increased toll of the eating disorders, anorexia nervosa and bulimia. Unfortunately, despite dramatic advances in the basic neurobiology of feeding, our understanding of the etiology of these conditions and our ability to intervene clinically remain limited.

9. Morton GJ, Meek TH, Schwartz MW (2014). "Neurobiology of food intake in health and disease". Nat. Rev. Neurosci. 15 (6): 367–378. doi:10.1038/nrn3745. PMC 4076116 . PMID 24840801. However, in normal individuals, body weight and body fat content are typically quite stable over time2,3 owing to a biological process termed 'energy homeostasis' that matches energy intake to expenditure over long periods of time. The energy homeostasis system comprises neurons in the mediobasal hypothalamus and other brain areas4 that are a part of a neurocircuit that regulates food intake in response to input from humoral signals that circulate at concentrations proportionate to body fat content4-6. ... An emerging concept in the neurobiology of food intake is that neurocircuits

exist that are normally inhibited, but when activated in response to emergent or stressful stimuli they can override the homeostatic control of energy balance. Understanding how these circuits interact with the energy homeostasis system is fundamental to understanding the control of food intake and may bear on the pathogenesis of disorders at both ends of the body weight spectrum.

10. Keith N. Frayn (2013). Metabolic Regulation: A Human Perspective. John Wiley & Sons. ISBN 1118685334.
11. Malenka RC, Nestler EJ, Hyman SE (2009). Sydor A, Brown RY, ed. Molecular Neuropharmacology: A Foundation for Clinical Neuroscience (2nd ed.). New York: McGraw-Hill Medical. pp. 179, 262–263. ISBN 9780071481274. Orexin neurons are regulated by peripheral mediators that carry information about energy balance, including glucose, leptin, and ghrelin. ... Accordingly, orexin plays a role in the regulation of energy homeostasis, reward, and perhaps more generally in emotion. ... The regulation of energy balance involves the exquisite coordination of food intake and energy expenditure. Experiments in the 1940s and 1950s showed that lesions of the lateral hypothalamus (LH) reduced food intake; hence, the normal role of this brain area is to stimulate feeding and decrease energy utilization. In contrast, lesions of the medial hypothalamus, especially the ventromedial nucleus (VMH) but also the PVN and dorsomedial hypothalamic nucleus (DMH), increased food intake; hence, the normal role of these regions is to suppress feeding and increase energy utilization. Yet discovery of the complex networks of neuropeptides and other neurotransmitters acting within the hypothalamus and other brain regions to regulate food intake and energy expenditure began in earnest in 1994 with the cloning of the leptin (ob, for obesity) gene. Indeed, there is now explosive interest in basic feeding mechanisms given the epidemic proportions of obesity in our society, and the increased toll of the eating disorders, anorexia nervosa and bulimia. Unfortunately, despite dramatic advances in the basic neurobiology of feeding, our understanding of the etiology of these conditions and our ability to intervene clinically remain limited.
12. Morton GJ, Meek TH, Schwartz MW (2014). "Neurobiology of food intake in health and disease". Nat. Rev. Neurosci. 15 (6): 367–378. doi:10.1038/nrn3745. PMC 4076116 . PMID 24840801. However, in normal individuals, body weight and body fat content are typically quite stable over time2,3 owing to a biological process termed 'energy homeostasis' that matches energy intake to expenditure over long periods of time. The energy homeostasis system comprises neurons in the mediobasal hypothalamus and other brain areas4 that are a part of a neurocircuit that regulates food intake in response to input from humoral signals that circulate at concentrations proportionate to body fat content4-6. ... An emerging concept in the neurobiology of food intake is that neurocircuits exist that are normally inhibited, but when activated in response to emergent or stressful stimuli they can override the homeostatic control of energy balance. Understanding how these circuits interact with the energy homeostasis system is fundamental to understanding the control of food intake and may bear on the pathogenesis of disorders at both ends of the body weight spectrum.

13. Farr OM, Li CS, Mantzoros CS (2016). "Central nervous system regulation of eating: Insights from human brain imaging". Metab. Clin. Exp. 65 (5): 699–713. doi:10.1016/j.metabol.2016.02.002. PMID 27085777.

14. Kevin G. Murphy & Stephen R. Bloom (December 14, 2006). "Gut hormones and the regulation of energy homeostasis". 444 (7121). Nature: 854–859. doi:10.1038/nature05484. PMID 17167473.

15. David Halliday, Robert Resnick, Jearl Walker, Fundamentals of physics, 9th edition,John Wiley & Sons, Inc., 2011, p. 485

16. M. D. Klok, S. Jakobsdottir and M. L. Drent. The role of leptin and ghrelin in the regulation of food intake and body weight in humans: a review. http://onlinelibrary.wiley.com/doi/10.1111/j.1467-789X.2006.00270.x/full

17. Anderson RM, Shanmuganayagam D, Weindruch R (2009). "Caloric restriction and aging: studies in mice and monkeys". Toxicol Pathol. 37 (1): 47–51. doi:10.1177/0192623308329476. PMID 19075044.

18. Rezzi S, Martin FP, Shanmuganayagam D, Colman RJ, Nicholson JK, Weindruch R (May 2009). "Metabolic shifts due to long-term caloric restriction revealed in nonhuman primates". Exp. Gerontol. 44 (5): 356–62. doi:10.1016/j.exger.2009.02.008. PMC 2822382 . PMID 19264119.

CHAPTER 9

Biological System

Chapter Contents

- **Introduction**
- **Overview**
- **Energy Homeostasis**
- **Energy**

A Biological system, (or organ system or body system) is a group of organs that work together to perform a certain task. Common systems, such as those present in mammals and other animals, seen in human anatomy, are those such as the circulatory system, the respiratory system, the nervous system, etc. Biological Systems, an academic journal provides an international, multi-disciplinary forum for the study of biological systems.

INTRODUCTION

A biological system is a complex network of biologically relevant entities. As biological organization spans several scales, examples of biological systems are populations of organisms, or on the organ- and tissue scale in mammals and other animals, the circulatory system, the respiratory system, the nervous system, etc.

On the micro to the nanoscopic scale, examples of biological systems are cells, organelles, macromolecular complexes and regulatory pathways.

A biological system is not to be confused with a living system, which is commonly referred to as life. For further information see e.g. definition of life or synthetic biology.

ORGAN AND TISSUE SYSTEMS

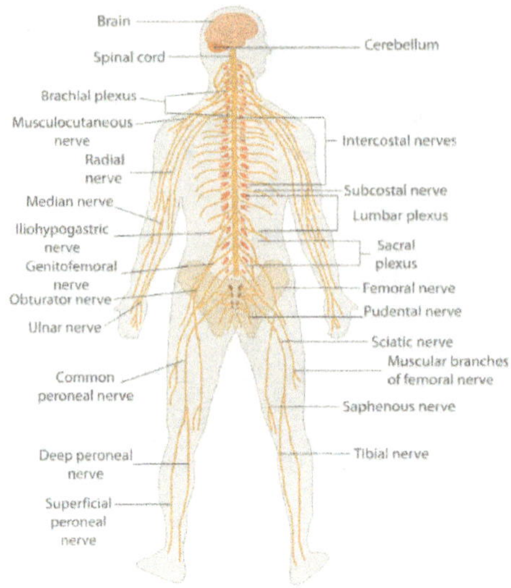

An example of a system: The brain, the cerebellum, the spinal cord, and the nerves are the four basic components of the nervous system.

These specific systems are widely studied in Human anatomy. *"Human" systems,* are also present in many other animals.

- Circulatory system: pumping and channeling blood to and from the body and lungs with heart, blood and blood vessels.
- Integumentary system: skin, hair, fat, and nails.
- Skeletal system: structural support and protection with bones, cartilage, ligaments and tendons.
- Reproductive system: the sex organs, such as ovaries, fallopian tubes, uterus, vagina, mammary glands, testes, vas deferens, seminal vesicles and prostate
- Digestive system: digestion and processing food with salivary glands, oesophagus, stomach, liver, gallbladder, pancreas, intestines, rectum and anus.
- Urinary system: kidneys, ureters, bladder and urethra involved in fluid balance, electrolyte balance and excretion of urine.
- Respiratory system: the organs used for breathing, the pharynx, larynx, bronchi, lungs and diaphragm.
- Endocrine system: communication within the body using hormones made by endocrine glands such as the hypothalamus, pituitary gland, pineal body or pineal gland, thyroid, parathyroid and adrenals, i.e., adrenal glands.
- Immune system: protects the organism from foreign bodies
    - Lymphatic system: structures involved in the transfer of lymph between tissues and the blood stream; includes the lymph and the nodes and vessels.

The lymphatic system includes functions including immune responses and development of antibodies.

- Muscular system: allows for manipulation of the environment, provides locomotion, maintains posture, and produces heat. Includes skeletal muscles, smooth muscles and cardiac muscle.
- Nervous system: collecting, transferring and processing information with brain, spinal cord and peripheral nervous system.

CIRCULATORY SYSTEM

The circulatory system, also called the cardiovascular system or the vascular system, is an organ system that permits blood to circulate and transport nutrients (such as amino acids and electrolytes), oxygen, carbon dioxide, hormones, and blood cells to and from the cells in the body to provide nourishment and help in fighting diseases, stabilize temperature and pH, and maintain homeostasis. The study of the blood flow is called hemodynamics. The study of the properties of the blood flow is called hemorheology.

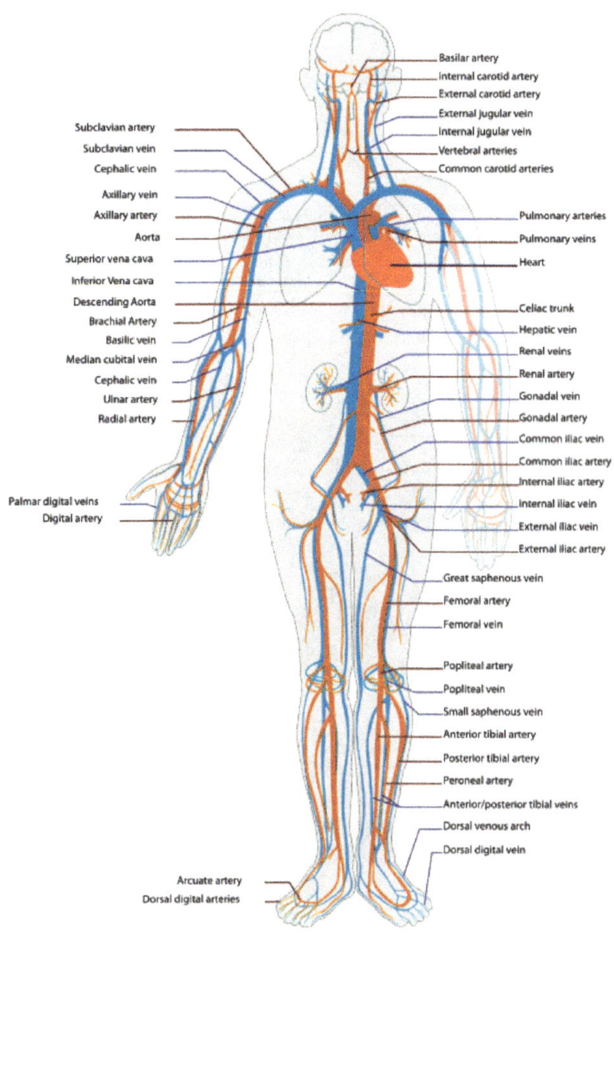

The circulatory system is often seen to comprise two separate systems: the cardiovascular system, which distributes blood, and the lymphatic system, which circulates lymph. The passage of lymph for example takes much longer than that of blood. Blood is a fluid consisting of plasma, red blood cells, white blood cells, and platelets that is circulated by the heart through the vertebrate vascular system, carrying oxygen and nutrients to and waste materials away from all body tissues. Lymph is essentially recycled excess blood plasma after it has been filtered from the interstitial fluid (between cells) and returned to the lymphatic system. The cardiovascular (from Latin words meaning "heart" and "vessel") system comprises the blood, heart, and blood vessels. The lymph, lymph nodes, and lymph vessels form the lymphatic system, which returns filtered blood plasma from the interstitial fluid (between cells) as lymph.

While humans, as well as other vertebrates, have a closed cardiovascular system (meaning that the blood never leaves the network of arteries, veins and capillaries), some invertebrate groups have an open cardiovascular system. The lymphatic system, on the other hand, is an open system providing an accessory route for excess interstitial fluid to be returned to the blood. The more primitive, diploblastic animal phyla lack circulatory systems.

Structure

Cardiovascular System

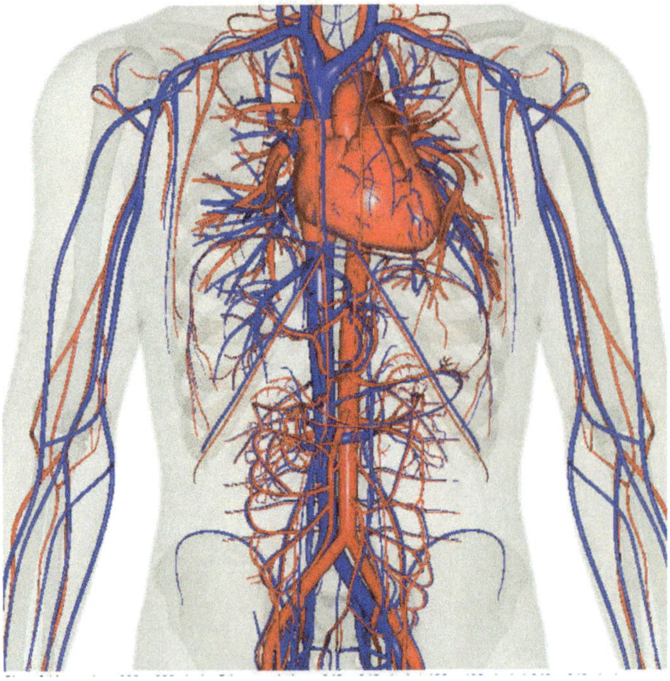

Depiction of the heart, major veins and arteries constructed from body scans.

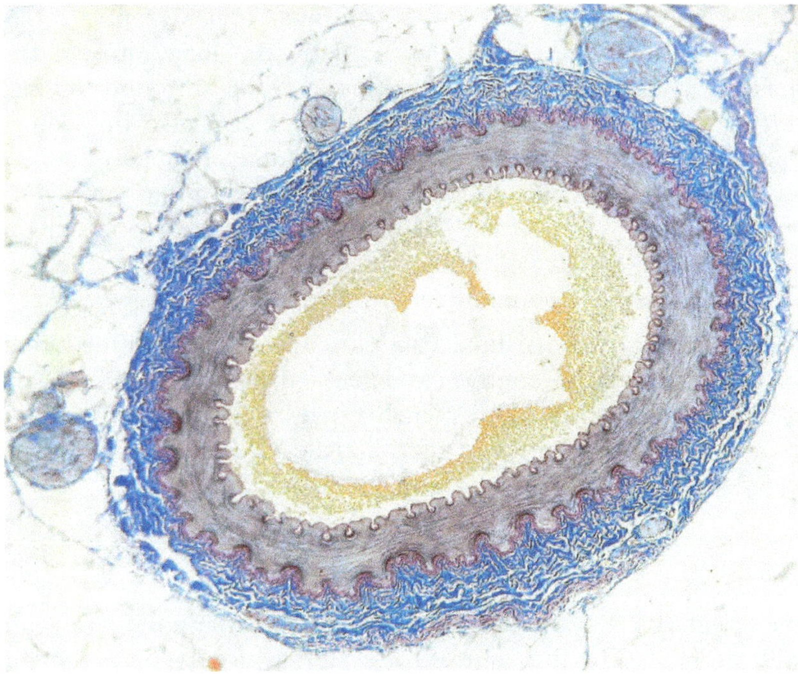

Cross section of a human artery

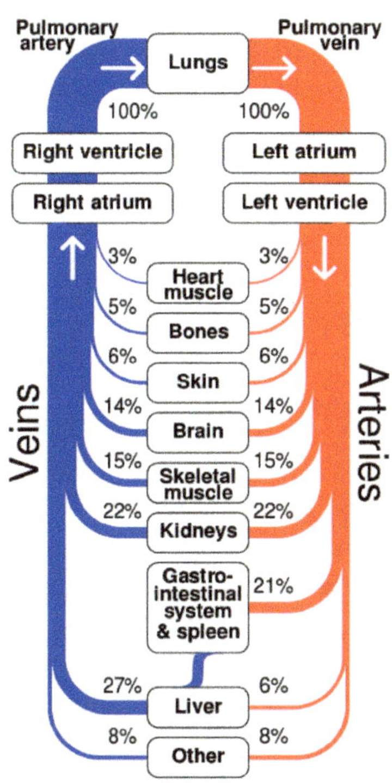

Relative percentages of cardiac output delivered to major organ systems

The essential components of the human cardiovascular system are the heart, blood and blood vessels. It includes the pulmonary circulation, a “loop” through the lungs where blood is oxygenated; and the systemic circulation, a “loop” through the rest of the body to provide oxygenated blood. The systemic circulation can also be seen to function in two parts–a *macrocirculation* and a microcirculation. An average adult contains five to six quarts (roughly 4.7 to 5.7 liters) of blood, accounting for approximately 7% of their total body weight. Blood consists of plasma, red blood cells, white blood cells, and platelets. Also, the digestive system works with the circulatory system to provide the nutrients the system needs to keep the heart pumping.

The cardiovascular systems of humans are closed, meaning that the blood never leaves the network of blood vessels. In contrast, oxygen and nutrients diffuse across the blood vessel layers and enter interstitial fluid, which carries oxygen and nutrients to the target cells, and carbon dioxide and wastes in the opposite direction. The other component of the circulatory system, the lymphatic system, is open.

Arteries

Oxygenated blood enters the systemic circulation when leaving the left ventricle, through the aortic semilunar valve. The first part of the systemic circulation is the aorta, a massive and thick-walled artery. The aorta arches and gives branches supplying the upper part of the body. after passing through the aortic opening of the diaphragm at the level of thoracic ten vertebra, it enters the abdomen. Later it descends down and supplies branches to abdomen, pelvis, perineum and the lower limbs. According to Difore’s Atlas of Histology the walls of aorta are elastic. This elasticity helps to maintain the blood pressure throughout the body. When the aorta receives almost 5 L of blood from the heart, it recoils and is responsible for pulsating B.P. Moreover, as aorta branches into smaller arteries, their elasticity goes on decreasing and their compliance goes on increasing.

Capillaries

Arteries branch into small passages called arterioles and then into the capillaries. The capillaries merge to bring blood into the venous system.

Veins

After their passage through body tissues, capillaries merge once again into venules, which continue to merge into veins. The venous system finally coalesces into two major veins: the superior vena cava (roughly speaking draining the areas above the heart) and the inferior vena cava (roughly speaking from areas below the heart). These two great vessels empty into the right atrium of the heart.

Coronary Vessels

The heart itself is supplied with oxygen and nutrients through a small “loop” of the systemic circulation and derives very little from the blood contained within the four chambers.

Portal Veins

The general rule is that arteries from the heart branch out into capillaries, which collect into veins leading back to the heart. Portal veins are a slight exception to this. In humans the only significant example is the hepatic portal vein which combines from capillaries around the gut where the blood absorbs the various products of digestion; rather than leading directly back to the heart, the hepatic portal vein branches into a second capillary system in the liver.

Heart

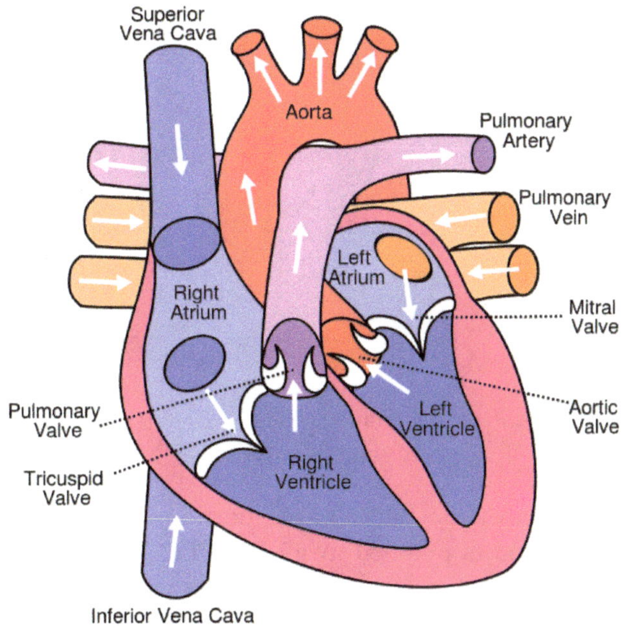

View from the front

The heart pumps oxygenated blood to the body and deoxygenated blood to the lungs. In the human heart there is one atrium and one ventricle for each circulation, and with both a systemic and a pulmonary circulation there are four chambers in total: left atrium, left ventricle, right atrium and right ventricle. The right atrium is the upper chamber of the right side of the heart. The blood that is returned to the right atrium is deoxygenated (poor in oxygen) and passed into the right ventricle to be pumped through the pulmonary artery to the lungs for re-oxygenation and removal of carbon dioxide. The left atrium receives newly oxygenated blood from the lungs as well as the pulmonary vein which is passed into the strong left ventricle to be pumped through the aorta to the different organs of the body.

The coronary circulation system provides a blood supply to the heart muscle itself. The coronary circulation begins near the origin of the aorta by two arteries: the right coronary artery and the left coronary artery. After nourishing the heart muscle, blood returns through the coronary veins into the coronary sinus and from this one into the right atrium. Back flow of blood through its opening during atrial systole is prevented by the Thebesian valve. The smallest cardiac veins drain directly into the heart chambers.

Lungs

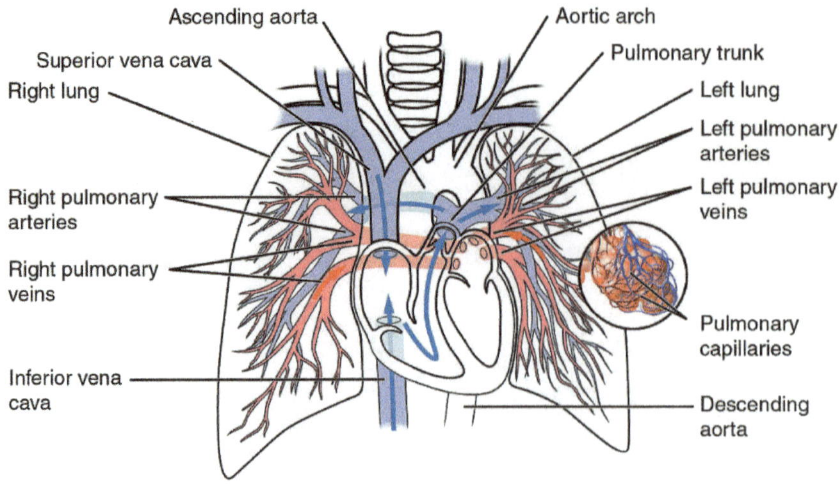

The pulmonary circulation as it passes from the heart. Showing both the pulmonary artery and bronchial arteries.

The circulatory system of the lungs is the portion of the cardiovascular system in which oxygen-depleted blood is pumped away from the heart, via the pulmonary artery, to the lungs and returned, oxygenated, to the heart via the pulmonary vein.

Oxygen deprived blood from the superior and inferior vena cava enters the right atrium of the heart and flows through the tricuspid valve (right atrioventricular valve) into the right ventricle, from which it is then pumped through the pulmonary semilunar valve into the pulmonary artery to the lungs. Gas exchange occurs in the lungs, whereby CO_2 is released from the blood, and oxygen is absorbed. The pulmonary vein returns the now oxygen-rich blood to the left atrium.

A separate system known as the bronchial circulation supplies blood to the tissue of the larger airways of the lung.

Systemic Circulation

The systemic circulation is the circulation of the blood to all parts of the body except the lungs. Systemic circulation is the portion of the cardiovascular system which transports oxygenated blood away from the heart through the aorta from the left ventricle where the blood has been previously deposited from pulmonary circulation, to the rest of the body, and returns oxygen-depleted blood back to the heart.

Brain

The brain has a dual blood supply that comes from arteries at its front and back. These are called the "anterior" and "posterior" circulation respectively. The anterior circulation arises from the internal carotid arteries and supplies the front of the brain. The posterior circulation arises from the vertebral arteries, and supplies the back of the brain and brainstem. The circulation from the front and the back join together (anastomise) at the Circle of Willis.

Kidneys

The renal circulation receives around 20% of the cardiac output. It branches from the abdominal aorta and returns blood to the ascending vena cava. It is the blood supply to the kidneys, and contains many specialized blood vessels.

Lymphatic System

The lymphatic system is part of the circulatory system. It is a network of lymphatic vessels and lymph capillaries, lymph nodes and organs, and lymphatic tissues and circulating lymph. One of its major functions is to carry the lymph, draining and returning interstitial fluid back towards the heart for return to the cardiovascular system, by emptying into the lymphatic ducts. Its other main function is in the adaptive immune system.

Physiology

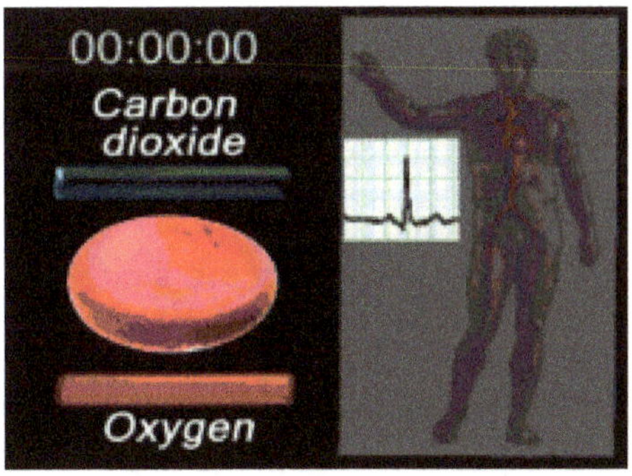

Typical human red blood cell cycle in the circulatory system. This animation occurs at a faster rate (~20 seconds of the average 60-second cycle) and shows the red blood cell deforming as it enters capillaries, as well as the bars changing color as the cell alternates in states of oxygenation along the circulatory system.

About 98.5% of the oxygen in a sample of arterial blood in a healthy human, breathing air at sea-level pressure, is chemically combined with hemoglobin molecules. About 1.5% is physically dissolved in the other blood liquids and not connected to hemoglobin. The hemoglobin molecule is the primary transporter of oxygen in mammals and many other species.

Development

The development of the circulatory system starts with vasculogenesis in the embryo. The human arterial and venous systems develop from different areas in the embryo. The arterial system develops mainly from the aortic arches, six pairs of arches which develop on the upper part of the embryo. The venous system arises from three bilateral veins during weeks 4 – 8 of embryogenesis. Fetal circulation begins within the 8th week of development. Fetal circulation does not include the lungs, which are bypassed via the truncus arteriosus. Before birth the fetus obtains oxygen (and nutrients) from the mother through the placenta and the umbilical cord.

Arterial Development

The human arterial system originates from the aortic arches and from the dorsal aortae starting from week 4 of embryonic life. The first and second aortic arches regress and forms only the maxillary arteries and stapedial arteries respectively. The arterial system itself arises from aortic arches 3, 4 and 6 (aortic arch 5 completely regresses).

The dorsal aortae, present on the dorsal side of the embryo, are initially present on both sides of the embryo. They later fuse to form the basis for the aorta itself. Approximately thirty smaller arteries branch from this at the back and sides. These branches form the intercostal arteries, arteries of the arms and legs, lumbar arteries and the lateral sacral arteries. Branches to the sides of the aorta will form the definitive renal, suprarenal and gonadal arteries. Finally, branches at the front of the aorta consist of the vitelline arteries and umbilical arteries. The vitelline arteries form the celiac, superior and inferior mesenteric arteries of the gastrointestinal tract. After birth, the umbilical arteries will form the internal iliac arteries.

Venous Development

The human venous system develops mainly from the vitelline veins, the umbilical veins and the cardinal veins, all of which empty into the sinus venosus.

Clinical Significance

Many diseases affect the circulatory system. This includes cardiovascular disease, affecting the cardiovascular system, and lymphatic disease affecting the lymphatic system. Cardiologists are medical professionals which specialise in the heart, and cardiothoracic surgeons specialise in operating on the heart and its surrounding areas. Vascular surgeons focus on other parts of the circulatory system.

Cardiovascular Disease

Diseases affecting the cardiovascular system are called cardiovascular disease.

Many of these diseases are called "lifestyle diseases" because they develop over time and are related to a person's exercise habits, diet, whether they smoke, and other lifestyle choices a person makes. Atherosclerosis is the precursor to many of these diseases. It is where small atheromatous plaques build up in the walls of medium and large arteries. This may eventually grow or rupture to occlude the arteries. It is also a risk factor for acute coronary syndromes, which are diseases which are characterised by a sudden deficit of oxygenated blood to the heart tissue. Atherosclerosis is also associated with problems such as aneurysm formation or splitting ("dissection") of arteries.

Another major cardiovascular disease involves the creation of a clot, called a "thrombus". These can originate in veins or arteries. Deep venous thrombosis, which mostly occurs in the legs, is one cause of clots in the veins of the legs, particularly when a person has been stationary for a long time. These clots may embolise, meaning travel to another location in the body. The results of this may include pulmonary embolus, transient ischaemic attacks, or stroke.

Cardiovascular diseases may also be congenital in nature, such as heart defects or persistent fetal circulation, where the circulatory changes that are supposed to happen after birth do not. Not all congenital changes to the circulatory system are associated with diseases, a large number are anatomical variations.

Measurement Techniques

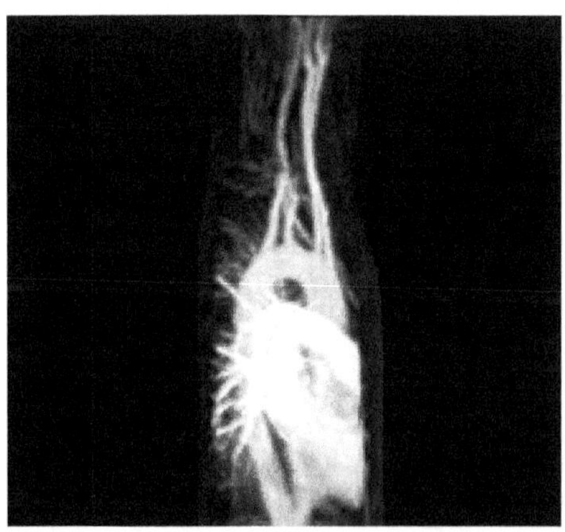

Magnetic resonance angiography of aberrant subclavian artery

The function and health of the circulatory system and its parts are measured in a variety of manual and automated ways. These include simple methods such as those that are part of the cardiovascular examination, including the taking of a person's pulse as an indicator of a person's heart rate, the taking of blood pressure through a sphygmomanometer or the use of a stethoscope to listen to the heart for murmurs which may indicate problems with the heart's valves. An electrocardiogram can also be used to evaluate the way in which electricity is conducted through the heart.

Other more invasive means can also be used. A cannula or catheter inserted into an artery may be used to measure pulse pressure or pulmonary wedge pressures. Angiography, which involves injecting a dye into an artery to visualise an arterial tree, can be used in the heart (coronary angiography) or brain. At the same time as the arteries are visualised, blockages or narrowings may be fixed through the insertion of stents, and active bleeds may be managed by the insertion of coils. An MRI may be used to image arteries, called an MRI angiogram. For evaluation of the blood supply to the lungs a CT pulmonary angiogram may be used.

Ultrasound can also be used, particularly to identify the health of blood vessels, and a Doppler ultrasound of the carotid arteries or Doppler ultrasound of the lower limbs can be used to evaluate for narrowing of the carotid arteries or thrombus formation in the legs, respectively.

Other Animals

Other Vertebrates

The circulatory systems of all vertebrates, as well as of annelids (for example, earthworms) and cephalopods (squids, octopuses and relatives) are *closed*, just as in humans. Still, the systems of fish, amphibians, reptiles, and birds show various stages of the evolution of the circulatory system.

In fish, the system has only one circuit, with the blood being pumped through the capillaries of the gills and on to the capillaries of the body tissues. This is known as *single cycle* circulation. The heart of fish is, therefore, only a single pump (consisting of two chambers).

In amphibians and most reptiles, a double circulatory system is used, but the heart is not always completely separated into two pumps. Amphibians have a three-chambered heart.

In reptiles, the ventricular septum of the heart is incomplete and the pulmonary artery is equipped with a sphincter muscle. This allows a second possible route of blood flow. Instead of blood flowing through the pulmonary artery to the lungs, the sphincter may be contracted to divert this blood flow through the incomplete ventricular septum into the left ventricle and out through the aorta. This means the blood flows from the capillaries to the heart and back to the capillaries instead of to the lungs. This process is useful to ectothermic (cold-blooded) animals in the regulation of their body temperature.

Birds, mammals, and crocodilians show complete separation of the heart into two pumps, for a total of four heart chambers; it is thought that the four-chambered heart of birds and crocodilians evolved independently from that of mammals.

Open Circulatory System

The open circulatory system is a system in which a fluid in a cavity called the hemocoel bathes the organs directly with oxygen and nutrients and there is no distinction between blood and interstitial fluid; this combined fluid is called hemolymph or haemolymph. Muscular movements by the animal during locomotion can facilitate hemolymph movement, but diverting flow from one area to another is limited. When the heart relaxes, blood is drawn back toward the heart through open-ended pores (ostia).

Hemolymph fills all of the interior hemocoel of the body and surrounds all cells. Hemolymph is composed of water, inorganicsalts (mostly sodium, chlorine, potassium, magnesium, and calcium), and organic compounds (mostly carbohydrates, proteins, and lipids). The primary oxygen transporter molecule is hemocyanin.

There are free-floating cells, the hemocytes, within the hemolymph. They play a role in the arthropod immune system.

Flatworms, such as this *Pseudoceros bifurcus*, lack specialized circulatory organs

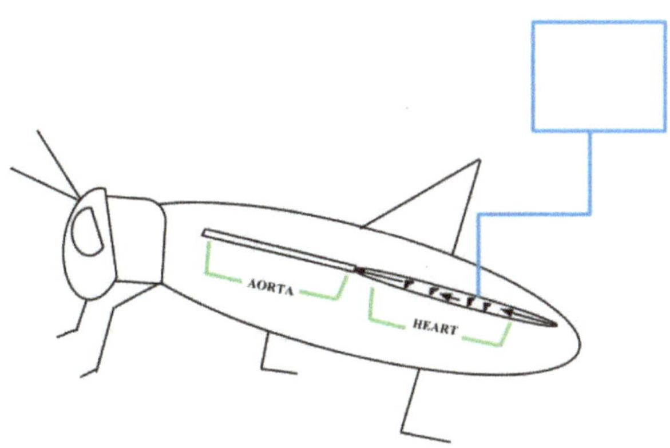

Above is a diagram of an open circulatory system. An open circulatory system is made up of a heart, vessels, and hemolymph. This diagram shows how the hemolymph, fluid present in most invertebrates that is equivalent to blood, is circulated throughout the body of a grasshopper. The hymolymph is first pumped through the heart, into the aorta, dispersed into the head and throughout the hemocoel, then back through the ostium that are located in the heart, where the process is repeated.

Absence of Circulatory System

Circulatory systems are absent in some animals, including flatworms (phylum Platyhelminthes). Their body cavity has no lining or enclosed fluid. Instead a muscular pharynx leads to an extensively branched digestive system that facilitates direct diffusion of nutrients to all cells. The flatworm's dorso-ventrally flattened body shape also restricts the distance of any cell from the digestive system or the exterior of the organism. Oxygen can diffuse from the surrounding water into the cells, and carbon dioxide can diffuse out. Consequently, every cell is able to obtain nutrients, water and oxygen without the need of a transport system.

Some animals, such as jellyfish, have more extensive branching from their gastrovascular cavity (which functions as both a place of digestion and a form of circulation), this branching allows for bodily fluids to reach the outer layers, since the digestion begins in the inner layers.

Human anatomical chart of blood vessels, with heart, lungs, liver and kidneys included. Other organs are numbered and arranged around it. Before cutting out the figures on this page, Vesalius suggests that readers glue the page onto parchment and gives instructions on how to assemble the pieces and paste the multilayered figure onto a base "muscle man" illustration.

The earliest known writings on the circulatory system are found in the Ebers Papyrus (16th century BCE), an ancient Egyptian medical papyrus containing over 700 prescriptions and remedies, both physical and spiritual. In the papyrus, it acknowledges the connection of the heart to the arteries. The Egyptians thought air came in through the mouth and into the lungs and heart. From the heart, the air travelled to every member through the arteries. Although this concept of the circulatory system is only partially correct, it represents one of the earliest accounts of scientific thought.

In the 6th century BCE, the knowledge of circulation of vital fluids through the body was known to the Ayurvedic physician Sushruta in ancient India. He also seems to have possessed knowledge of the arteries, described as 'channels' by Dwivedi & Dwivedi (2007). The valves of the heart were discovered by a physician of the Hippocratean school around the 4th century BCE. However their function was not properly understood then. Because blood pools in the veins after death, arteries look empty. Ancient anatomists assumed they were filled with air and that they were for transport of air.

The Greek physician, Herophilus, distinguished veins from arteries but thought that the pulse was a property of arteries themselves. Greek anatomist Erasistratus observed that arteries that were cut during life bleed. He ascribed the fact to the phenomenon that air escaping from an artery is replaced with blood that entered by very small vessels between veins and arteries. Thus he apparently postulated capillaries but with reversed flow of blood.

In 2nd century AD Rome, the Greek physician Galen knew that blood vessels carried blood and identified venous (dark red) and arterial (brighter and thinner) blood, each with distinct and separate functions. Growth and energy were derived from venous blood created in the liver from chyle, while arterial blood gave vitality by containing pneuma (air) and originated in the heart. Blood flowed from both creating organs to all parts of the body where it was consumed and there was no return of blood to the heart or liver. The heart did not pump blood around, the heart›s motion sucked blood in during diastole and the blood moved by the pulsation of the arteries themselves.

Galen believed that the arterial blood was created by venous blood passing from the left ventricle to the right by passing through ‹pores› in the interventricular septum, air passed from the lungs via the pulmonary artery to the left side of the heart. As the arterial blood was created ‹sooty› vapors were created and passed to the lungs also via the pulmonary artery to be exhaled.

In 1025, *The Canon of Medicine* by the Persian physician, Avicenna, "erroneously accepted the Greek notion regarding the existence of a hole in the ventricular septum by which the blood traveled between the ventricles." Despite this, Avicenna "correctly wrote on the cardiac cycles and valvular function", and "had a vision of blood circulation" in his *Treatise on Pulse*. While also refining Galen's erroneous theory of the pulse, Avicenna provided the first correct explanation of pulsation: "Every beat of the pulse comprises two movements and two pauses. Thus, expansion : pause : contraction : pause. [...] The pulse is a movement in the heart and arteries ... which takes the form of alternate expansion and contraction."

In 1242, the Arabian physician, Ibn al-Nafis, became the first person to accurately describe the process of pulmonary circulation, for which he is sometimes considered the father of circulatory physiology. Ibn al-Nafis stated in his *Commentary on Anatomy in Avicenna's Canon*:

"...the blood from the right chamber of the heart must arrive at the left chamber but there is no direct pathway between them. The thick septum of the heart is not perforated and does not have visible pores as some people thought or invisible pores as Galen thought. The blood from the right chamber must flow through the vena arteriosa (pulmonary artery) to the lungs, spread through its substances, be mingled there with air, pass through the arteria venosa (pulmonary vein) to reach the left chamber of the heart and there form the vital spirit..."

In addition, Ibn al-Nafis had an insight into what would become a larger theory of the capillary circulation. He stated that "there must be small communications or pores (*manafidh* in Arabic) between the pulmonary artery and vein," a prediction that preceded the discovery of the capillary system by more than 400 years. Ibn al-Nafis› theory, however, was confined to blood transit in the lungs and did not extend to the entire body.

Michael Servetus was the first European to describe the function of pulmonary circulation, although his achievement was not widely recognized at the time, for a few reasons. He firstly described it in the "Manuscript of Paris" (near 1546), but this work was never published. And later he published this description, but in a theological treatise, *Christianismi Restitutio*, not in a book on medicine. Only three copies of the book survived but these remained hidden for decades, the rest were burned shortly after its publication in 1553 because of persecution of Servetus by religious authorities.

Better known discovery of pulmonary circulation was by Vesalius's successor at Padua, Realdo Colombo, in 1559.

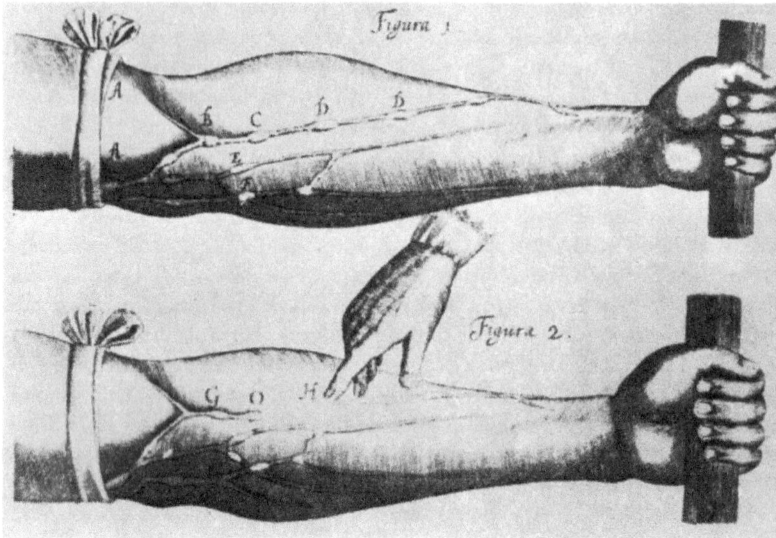

Image of veins from William Harvey's *Exercitatio Anatomica de Motu Cordis et Sanguinis in Animalibus*

Finally, William Harvey, a pupil of Hieronymus Fabricius (who had earlier described the valves of the veins without recognizing their function), performed a sequence of experiments, and published *Exercitatio Anatomica de Motu Cordis et Sanguinis in Animalibus* in 1628, which "demonstrated that there had to be a direct connection between the venous and arterial systems throughout the body, and not just the lungs. Most importantly, he argued that the beat of the heart produced a continuous circulation of blood through minute connections at the extremities of the body. This is a conceptual leap that was quite different from Ibn al-Nafis› refinement of the anatomy and bloodflow in the heart and lungs." This work, with its essentially correct exposition, slowly convinced the medical world. However, Harvey was not able to identify the capillary system connecting arteries and veins; these were later discovered by Marcello Malpighi in 1661.

In 1956, André Frédéric Cournand, Werner Forssmann and Dickinson W. Richards were awarded the Nobel Prize in Medicine "for their discoveries concerning heart catheterization and pathological changes in the circulatory system."

In the 1970s, Diana McSherry developed computer-based systems to create images of the circulatory system and heart without the need for surgery.

INTEGUMENTARY SYSTEM

The integumentary system is the organ system that protects the body from various kinds of damage, such as loss of water or abrasion from outside. The system comprises the skin and its appendages (including hair, scales, feathers, hooves, and nails). The integumentary system has a variety of functions; it may serve to waterproof, cushion, and protect the deeper tissues, excrete wastes, and regulate temperature, and is the attachment site for sensory receptors to detect pain, sensation, pressure, and temperature. In most terrestrial vertebrates with significant exposure to sunlight, the integumentary system also provides for vitamin D synthesis.

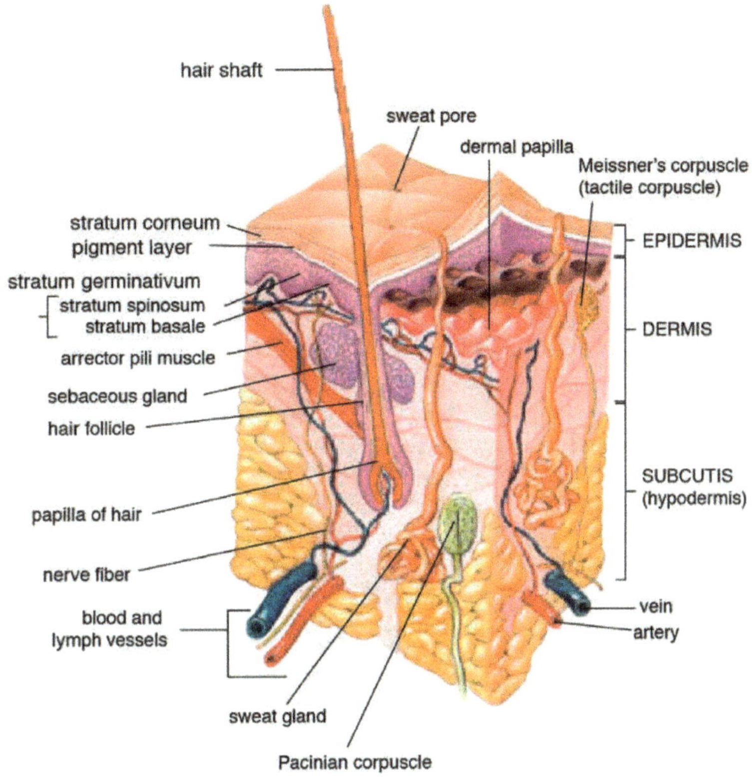

Cross-section of all skin layers

Skin

The skin is the largest organ in the body. In humans, it accounts for about 12 to 15 percent of total body weight and covers 1.5-2m^2 of surface area. It distinguishes, separates, and protects the organism from its surroundings. Small-bodied invertebrates of aquatic or continually moist habitats respire using the outer layer (integument). This gas exchange system, where gases simply diffuse into and out of the interstitial fluid, is called integumentary exchange.

The human skin (integument) is composed of a minimum of two major layers of tissue: the epidermis and dermis. (The hypodermis or subcutaneous layer is not part of the skin.) The epidermis forms the outermost layer, providing the initial barrier to the external environment. Beneath this, the dermis comprises two sections, the papillary and reticular layers, and contains connective tissues, vessels, glands, follicles, hair roots, sensory nerve endings, and muscular tissue. The deepest layer is the hypodermis, which is primarily made up of adipose tissue. Substantial collagen bundles anchor the dermis to the hypodermis in a way that permits most areas of the skin to move freely over the, deeper tissue layers.

Epidermis

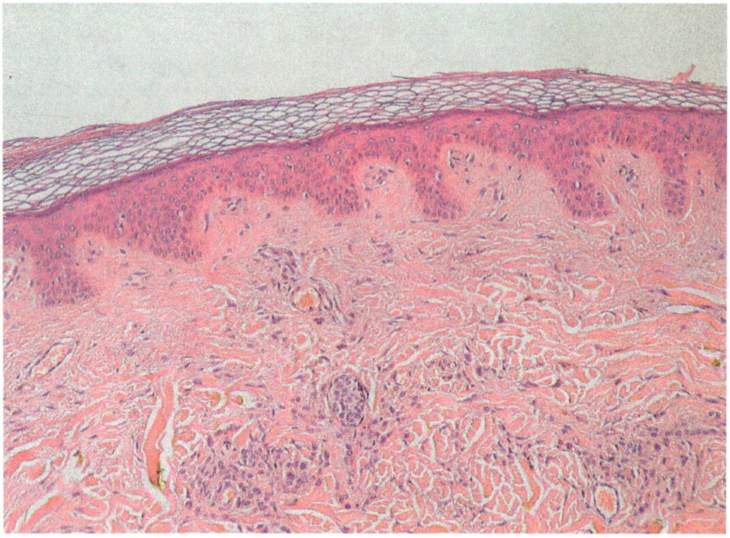

Epidermis and dermis of human skin

This is the top layer of skin made up of epithelial cells. It does not contain blood vessels. Its main function is protection, absorption of nutrients, and homeostasis. In structure, it consists of a keratinized stratified squamous epithelium comprising four types of cells: keratinocytes, melanocytes, Merkel cells, and Langerhans' cells. The major cell of the epidermis is the keratinocyte, which produces keratin. Keratin is a fibrous protein that aids in protection. Keratin is also a waterproofing protein. Millions of dead keratinocytes rub off daily. The majority of the skin on the body is keratinized. The only skin on the body that is non-keratinized is the lining of skin on the inside of the mouth. Non-keratinized cells allow water to "stay" atop the structure.

The protein keratin stiffens epidermal tissue to form fingernails. Nails grow from a thin area called the nail matrix; growth of nails is 1 mm per week on average. The lunula is the crescent-shape area at the base of the nail, this is a lighter color as it mixes with the matrix cells.

Dermis

The dermis is the middle layer of skin, composed of dense irregular connective tissue and areolar connective tissue such as a collagen with elastin arranged in a diffusely bundled and woven pattern. The dermis has two layers. One is the papillary layer which is the superficial layer and consists of the areolar connective tissue. The other is the reticular layer which is the deep layer of the dermis and consists of the dense irregular connective tissue. These layers serve to give elasticity to the integument, allowing stretching and conferring flexibility, while also resisting distortions, wrinkling, and sagging. The dermal layer provides a site for the endings of blood vessels and nerves. Many chromatophores are also stored in this layer, as are the bases of integumental structures such as hair, feathers, and glands.

Hypodermis

The hypodermis, otherwise known as the subcutaneous layer, is a layer beneath the skin. It invaginates into the dermis and is attached to the latter, immediately above it, by collagen and elastin fibres. It is essentially composed of a type of cell known as adipocytes specialised in accumulating and storing fats. These cells are grouped together in lobules separated by connective tissue.

The hypodermis acts as an energy reserve. The fats contained in the adipocytes can be put back into circulation, via the venous route, during intense effort or when there is a lack of energy providing substances, and are then transformed into energy. The hypodermis participates, passively at least, in thermoregulation since fat is a heat insulator.

Functions

The integumentary system has multiple roles in homeostasis. All body systems work in an interconnected manner to maintain the internal conditions essential to the function of the body. The skin has an important job of protecting the body and acts as the body's first line of defense against infection, temperature change, and other challenges to homeostasis. Functions include:

- Protect the body's internal living tissues and organs
- Protect against invasion by infectious organisms
- Protect the body from dehydration
- Protect the body against abrupt changes in temperature, maintain homeostasis
- Help excrete waste materials through perspiration
- Act as a receptor for touch, pressure, pain, heat, and cold (see Somatosensory system)
- Protect the body against sunburns by secreting melanin
- Generate vitamin D through exposure to ultraviolet light
- Store water, fat, glucose, vitamin D
- Maintenance of the body form
- Formation of new cells from stratum germanium to repair minor injuries
- Protect from UV rays.

Clinical Significance

Possible diseases and injuries to the human integumentary system include:

- Rash
- Yeast
- Athlete's foot
- Infection
- Sunburn
- Skin cancer
- Albinism
- Acne
- Herpes
- Herpes labialis, commonly called cold sores
- Impetigo
- Rubella
- Cancer
- Psoriasis
- Rabies
- Rosacea

SKELETON

The skeleton is the body part that forms the supporting structure of an organism. There are several different skeletal types: the exoskeleton, which is the stable outer shell of an organism, the endoskeleton, which forms the support structure inside the body, the hydroskeleton, and the cytoskeleton.

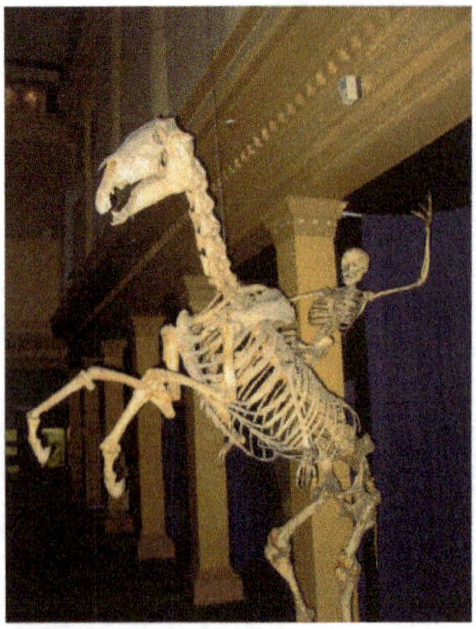

The skeletons of a human and a horse placed in a display in the Australian Museum, Sydney.

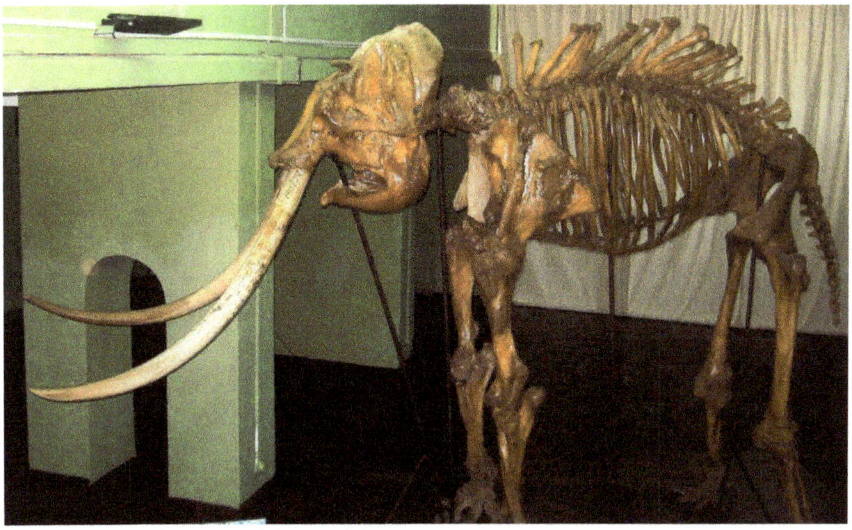

Heiyantuduwa Raja (elephant) Skeleton at National Museum of Colombo Sri Lanka

Types of Skeletons

There are two major types of skeletons: solid and fluid. Solid skeletons can be internal, called an endoskeleton, or external, called an exoskeleton, and may be further classified as pliant (elastic/movable) or rigid (hard/non-movable). Fluid skeletons are always internal.

Exoskeleton

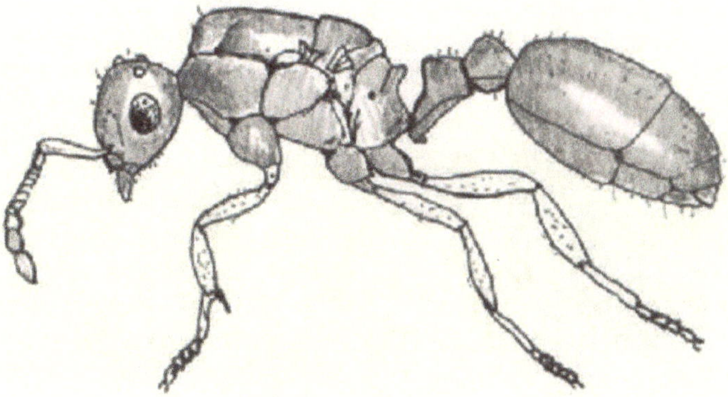

Exoskeleton of an ant

Exoskeletons are external, and are found in many invertebrates; they enclose and protect the soft tissues and organs of the body. Some kinds of exoskeletons undergo periodic moulting as the animal grows, as is the case in many arthropods including insects and crustaceans. The exoskeleton of insects is not only a form of protection, but also serves as a surface for muscle attachment, as a watertight protection against drying, and as a sense organ to interact with the environment. The shell of mollusks also performs all of the same functions, except that in most cases it does not contain sense organs.

An external skeleton can be quite heavy in relation to the overall mass of an animal, so on land, organisms that have an exoskeleton are mostly relatively small. Somewhat larger aquatic animals can support an exoskeleton because weight is less of a consideration underwater. The southern giant clam, a species of extremely large saltwater clam in the Pacific Ocean, has a shell that is massive in both size and weight. *Syrinx aruanus* is a species of sea snail with a very large shell.

Endoskeleton

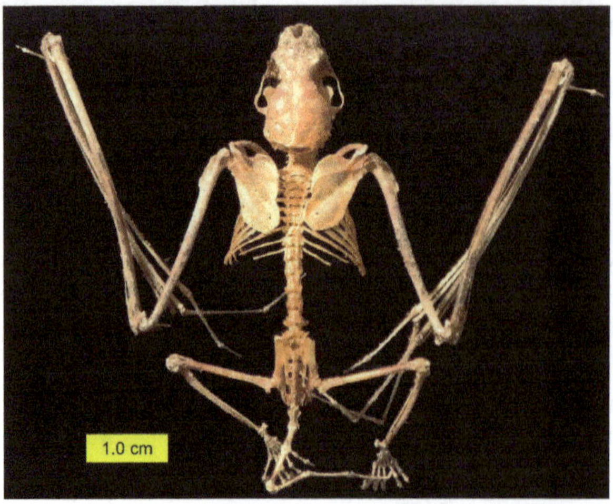

Endoskeleton of a bat.

A beaver skeleton on display at The Museum of Osteology, Oklahoma City, Oklahoma.

The endoskeleton is the internal support structure of an animal, composed of mineralized tissue and is typical of vertebrates. Endoskeletons vary in complexity from functioning purely for support (as in the case of sponges), to serving as an attachment site for muscles and a mechanism for transmitting muscular forces. A true endoskeleton is derived from mesodermal tissue. Such a skeleton is present in echinoderms and chordates.

Pliant Skeletons

Pliant skeletons are capable of movement; thus, when stress is applied to the skeletal structure, it deforms and then reverts to its original shape. This skeletal structure is used in some invertebrates, for instance in the hinge of bivalve shells or the mesoglea of cnidarians such as jellyfish. Pliant skeletons are beneficial because only muscle contractions are needed to bend the skeleton; upon muscle relaxation, the skeleton will return to its original shape. Cartilage is one material that a pliant skeleton may be composed of, but most pliant skeletons are formed from a mixture of proteins, polysaccharides, and water. For additional structure or protection, pliant skeletons may be supported by rigid skeletons. Organisms that have pliant skeletons typically live in water, which supports body structure in the absence of a rigid skeleton.

Rigid Skeletons

Rigid skeletons are not capable of movement when stressed, creating a strong support system most common in terrestrial animals. Such a skeleton type used by animals that live in water are more for protection (such as barnacle and snail shells) or for fast-moving animals that require additional support of musculature needed for swimming through water. Rigid skeletons are formed from materials including chitin (in arthropods), calcium compounds such as calcium carbonate (in stony corals and mollusks) and silicate (for diatoms and radiolarians).

Cytoskeleton

The cytoskeleton (gr. *kytos* = cell) is used to stabilize and preserve the form of the cells. It is a dynamic structure that maintains cell shape, protects the cell, enables cellular motion (using structures such as flagella, cilia and lamellipodia), and plays important roles in both intracellular transport (the movement of vesicles and organelles, for example) and cellular division.

Fluid Skeletons

Hydrostatic Skeleton (Hydroskeleton)

A hydrostatic skeleton is a semi-rigid, soft tissue structure filled with liquid under pressure, surrounded by muscles. Longitudinal and circular muscles around their body sectors allow movement by alternate lengthening and contractions along their lengths. A common example of this is earthworm.

Organisms with Skeletons

Invertebrates

The endoskeletons of echinoderms and some other soft-bodied invertebrates such as jellyfish and earthworms are also termed hydrostatic; a body cavity the coelom is filled with coelomic fluid and the pressure from this fluid acts together with the surrounding muscles to change the organism's shape and produce movement.

Sponges

The skeleton of sponges consists of microscopic calcareous or silicious spicules. The demosponges include 90% of all species of sponges. Their "skeletons" are made of spicules consisting of fibers of the protein spongin, the mineral silica, or both. Where spicules of silica are present, they have a different shape from those in the otherwise similar glass sponges.

Echinoderms

The skeleton of the echinoderms, which include, among other things, the starfish, is composed of calcite and a small amount of magnesium oxide. It lies below the epidermis in the mesoderm and is within cell clusters of frame-forming cells. This structure formed is porous and therefore firm and at the same time light. It coalesces into small calcareous ossicles (bony plates), which can grow in all directions and thus can replace the loss of a body part. Connected by joints, the individual skeletal parts can be moved by the muscles.

Vertebrates

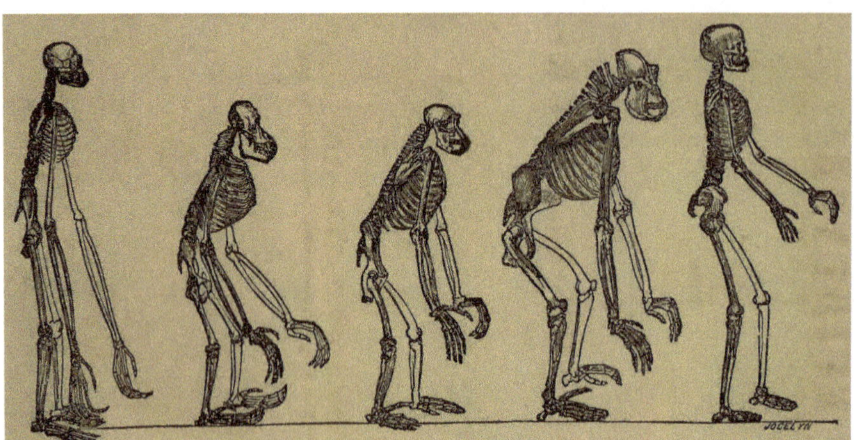

Pithecometra: From Thomas Huxley's 1863 *Evidence as to Man's Place in Nature*, the compared skeletons of apes to humans.

In most vertebrates, the main skeletal component is referred to as bone. Another important component is cartilage which in mammals is found mainly in the joint areas. In other animals, such as the cartilaginous fishes, which include the sharks, the skeleton is composed entirely of cartilage. The segmental pattern of the skeleton is present in all vertebrates (mammals, birds, fish, reptiles and amphibians) with basic units being repeated. This segmental pattern is particularly evident in the vertebral column and the ribcage.

Bones in addition to supporting the body also serve, at the cellular level, as calcium and phosphate storage.

Fish

The skeleton, which forms the support structure inside the fish is either made of cartilage as in the (Chondrichthyes), or bones as in the (Osteichthyes). The main skeletal element is the vertebral column, composed of articulating vertebrae which are lightweight yet strong. The ribs attach to the spine and there are no limbs or limb girdles. They are supported only by the muscles. The main external features of the fish, the fins, are composed of either bony or soft spines called rays, which with the exception of the caudal fin (tail fin), have no direct connection with the spine. They are supported by the muscles which compose the main part of the trunk.

Birds

The bird skeleton is highly adapted for flight. It is extremely lightweight, yet still strong enough to withstand the stresses of taking off, flying, and landing. One key adaptation is the fusing of bones into single ossifications, such as the pygostyle. Because of this, birds usually have a smaller number of bones than other terrestrial vertebrates. Birds also lack teeth or even a true jaw, instead having evolved a beak, which is far more lightweight. The beaks of many baby birds have a projection called an egg tooth, which facilitates their exit from the amniotic egg.

Marine Mammals

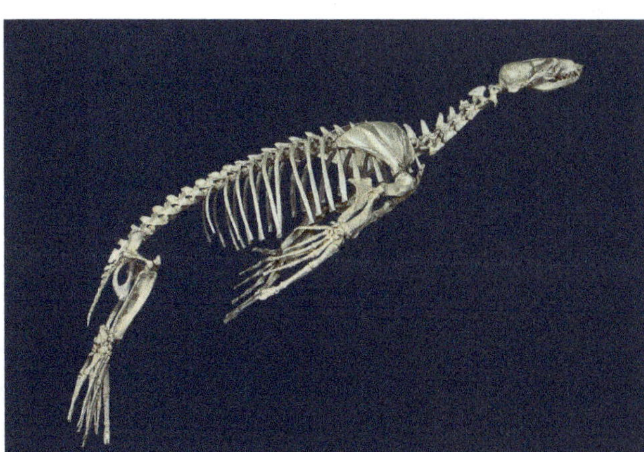

Californian sea lion

To facilitate the movement of marine mammals in water, the hind legs were either lost altogether, as in the whales and manatees, or united in a single tail fin as in the pinnipeds (seals). In the whale, the cervical vertebrae are typically fused, an adaptation trading flexibility for stability during swimming.

Humans

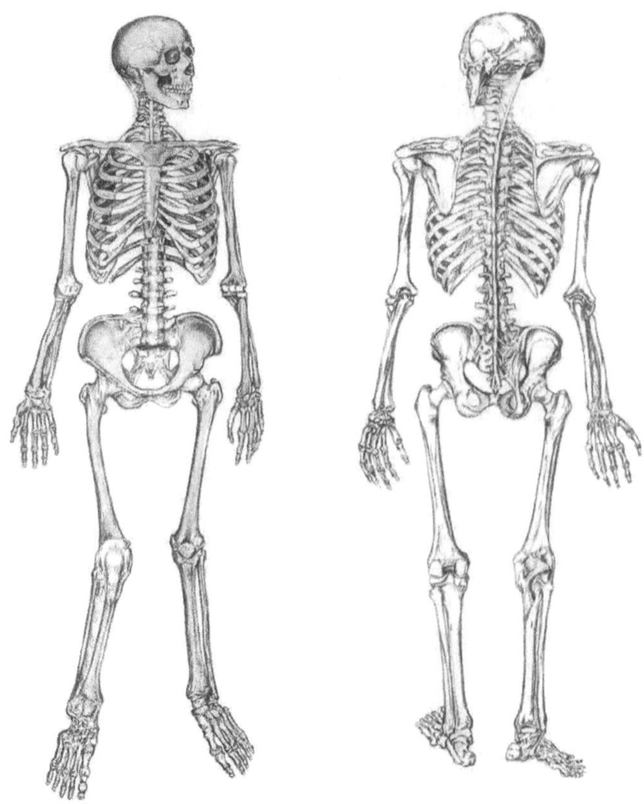

Study of Skeletons, c. 1510, by Leonardo da Vinci

The human skeleton consists of both fused and individual bones supported and supplemented by ligaments, tendons, muscles and cartilage. It serves as a scaffold which supports organs, anchors muscles, and protects organs such as the brain, lungs, heart and spinal cord. Although the teeth do not consist of tissue commonly found in bones, the teeth are usually considered as members of the skeletal system. The biggest bone in the body is the femur in the upper leg, and the smallest is the stapes bone in the middle ear. In an adult, the skeleton comprises around 14% of the total body weight, and half of this weight is water.

Fused bones include those of the pelvis and the cranium. Not all bones are interconnected directly: There are three bones in each middle ear called the ossicles that articulate only with each other. The hyoid bone, which is located in the neck and serves as the point of attachment for the tongue, does not articulate with any other bones in the body, being supported by muscles and ligaments.

There are 206 bones in the adult human skeleton, although this number depends on whether the pelvic bones (the hip bones on each side) are counted as one or three bones on each side (ilium, ischium, and pubis), whether the coccyx or tail bone is counted as one or four separate bones, and does not count the variable wormian bones between skull sutures. Similarly, the sacrum is usually counted as a single bone, rather than five fused vertebrae. There is also a variable number of small sesamoid bones, commonly found in tendons. The patella or kneecap on each side is an example of a larger sesamoid bone. The patellae are counted in the total, as they are constant. The number of bones varies between individuals and with age - newborn babies have over 270 bones some of which fuse together. These bones are organized into a longitudinal axis, the axial skeleton, to which the appendicular skeleton is attached.

The human skeleton takes 20 years before it is fully developed. In many animals, the skeleton bones contain marrow, which produces blood cells.

There exist several general differences between the male and female skeletons. The male skeleton, for example, is generally larger and heavier than the female skeleton. In the female skeleton, the bones of the skull are generally less angular. The female skeleton also has wider and shorter breastbone and slimmer wrists. There exist significant differences between the male and female pelvis which are related to the female›s pregnancy and childbirth capabilities. The female pelvis is wider and shallower than the male pelvis. Female pelvises also have an enlarged pelvic outlet and a wider and more circular pelvic inlet. The angle between the pubic bones is known to be sharper in males, which results in a more circular, narrower, and near heart-shaped pelvis.

REFERENCES

1. "circulatory system" at Dorland's Medical Dictionary
2. http://www.cancerresearch.uk
3. "cardiovascular system" at Dorland's Medical Dictionary
4. Sherwood, Lauralee (2011). Human Physiology: From Cells to Systems. Cengage Learning. pp. 401–. ISBN 978-1-133-10893-1.
5. Cardiovascular System at the US National Library of Medicine Medical Subject Headings (MeSH)
6. Pratt, Rebecca. "Cardiovascular System: Blood". AnatomyOne. Amirsys, Inc.
7. Guyton, Arthur; Hall, John (2000). Guyton Textbook of Medical Physiology (10 ed.). ISBN 072168677X.
8. National Institutes of Health. "What Are the Lungs?". nih.gov.
9. State University of New York (February 3, 2014). "The Circulatory System".

suny.edu.

10. Alberts, B.; Johnson, A.; Lewis, J.; Raff, M.; Roberts, K.; Walters, P. (2002). Molecular Biology of the Cell (4th ed.). New York and London: Garland Science. ISBN 0-8153-3218-1.
11. Whitaker, Kent (2001). "Fetal Circulation". Comprehensive Perinatal and Pediatric Respiratory Care. Delmar Thomson Learning. pp. 18–20. ISBN 978-0-7668-1373-1.
12. Wier LM, Steiner CA, Owens PL (April 17, 2015). "Surgeries in Hospital-Owned Outpatient Facilities, 2012". HCUP Statistical Brief #188. Rockville, MD: Agency for Healthcare Research and Quality.
13. "Crocodilian Hearts". National Center for Science Education. October 24, 2008. Retrieved October 3, 2015.
14. Bailey, Regina. "Circulatory System". biology.about.com.
15. Dwivedi, Girish & Dwivedi, Shridhar (2007). "History of Medicine: Sushruta – the Clinician – Teacher par Excellence"Archived October 10, 2008, at the Wayback Machine., Indian J Chest Dis Allied Sci Vol.49 pp.243-4, National Informatics Centre (Government of India).
16. Anatomy – History of anatomy. Scienceclarified.com. Retrieved 2013-09-15.
17. Shoja, M. M.; Tubbs, R. S.; Loukas, M.; Khalili, M.; Alakbarli, F.; Cohen-Gadol, A. A. (2009). "Vasovagal syncope in the Canon of Avicenna: The first mention of carotid artery hypersensitivity". International Journal of Cardiology. 134 (3): 297–301. doi:10.1016/j.ijcard.2009.02.035. PMID 19332359.
18. Hajar, Rachel (1999). "The Greco-Islamic Pulse". Heart Views. 1 (4): 136–140 .
19. Reflections, Chairman's (2004). "Traditional Medicine Among Gulf Arabs, Part II: Blood-letting". Heart Views. 5 (2): 74–85 .
20. West, J. B. (2008). "Ibn al-Nafis, the pulmonary circulation, and the Islamic Golden Age". Journal of Applied Physiology. 105 (6): 1877–1880. doi:10.1152/japplphysiol.91171.2008. PMC 2612469 . PMID 18845773.
21. Gonzalez Etxeberria, Patxi (2011) Amor a la verdad, el – vida y obra de Miguel servet [The love for truth. Life and work of Michael Servetus]. Navarro y Navarro, Zaragoza, collaboration with the Government of Navarra, Department of Institutional Relations and Education of the Government of Navarra. ISBN 8423532666 pp. 215–228 & 62nd illustration (XLVII)
22. Michael Servetus Research Study with graphical proof on the Manuscript of Paris and many other manuscripts and new works by Servetus
23. Pormann, Peter E. and Smith, E. Savage (2007) Medieval Islamic medicine Georgetown University, Washington DC, p. 48, ISBN 1589011619.
24. "The Nobel Prize in Physiology or Medicine 1956". Nobel Foundation. Retrieved 2007-07-28.
25. Wayne, Tiffany K. (2011). American women of science since 1900. Santa Barbara, Calif.: ABC-CLIO. pp. 677–678. ISBN 9781598841589.

26. "The Integumentary System". Encyclopedia.com. Retrieved 3 June 2013.
27. Integumentary System at the US National Library of Medicine Medical Subject Headings (MeSH)
28. Marieb, Elaine; Katja Hoehn (2007). Human Anatomy & Physiology (7th ed.). Pearson Benjamin Cummings. p. 142.
29. Martini & Nath: "Fundamentals of Anatomy & Physiology" 8th Edition, pp.158, Pearson Education, 2009
30. The Ageing Skin - Structure of Skin
31. Pratt, Rebecca. "Integument". AnatomyOne. Amirsys, Inc. Retrieved 2012-09-28.
32. Oxford Dictionary of English 2nd edition 2005
33. Barnes, Edward E.; Fox, Richard S.; Barnes, Robert D. (2003). Invertebrate zoology : a functional evolutionary approach (7. ed.). Belmont, Calif. [u.a.]: Thomson, Brooks/Cole. ISBN 0-03-025982-7.
34. Pechenik, Jan A. (2015). Biology of the Invertebrates (Seventh ed.). New York: McGraw-Hill Education. ISBN 978-0-07-352418-4.
35. Barnes, Robert D. (1982). Invertebrate Zoology. Philadelphia, PA: Holt-Saunders International. pp. 105–106. ISBN 0-03-056747-5.
36. "Beluga Whale". Yellowmagpie.com. 2012-06-27. Retrieved 2013-08-12.
37. "About Whales". Whalesalive.org.au. 2009-06-26. Retrieved 2013-08-12.
38. William W. Reynolds & William J. Karlotski (1977). "The Allometric Relationship of Skeleton Weight to Body Weight in Teleost Fishes: A Preliminary Comparison with Birds and Mammals". Copeia: 160–163. doi:10.2307/1443520.
39. Miller, Larry (2007-12-09). "We're Born With 270 Bones. As Adults We Have 206". Ground Report.
40. "How many bones does the human body contain?". Ask.yahoo.com. 2001-08-08. Retrieved 2010-03-04.
41. http://education.sdsc.edu/download/enrich/exploring_human.pdf
42. Tözeren, Aydın (2000). Human Body Dynamics: Classical Mechanics and Human Movement. Springer. pp. 6–10. ISBN 0-387-98801-7.
43. Balaban, Naomi (2008). The Handy Anatomy Answer Book. Visible Ink Press. p. 61. ISBN 978-1-57859-190-9.
44. Stein, Lisa (2007). Body The Complete Human: How It Grows, How It Works, And How to Keep It Healthy And Strong. National Geographic Society. p. 73. ISBN 978-1-4262-0128-8.
45. Steele, D. Gentry; Claud A. Bramblett (1988). The Anatomy and Biology of the Human Skeleton. Texas A&M University Press. p. 4. ISBN 0-89096-300-2.
46. Schmiedeler, Edgar; Mary Rosa McDonough (1934). Parent and Child: An Introductory Study of Parent Education. D. Appleton-Century. p. 31.

Index

L

M

N

O

P

R

S

T

U

X